# Analysis
# of Industrial
# Wastewaters

# Analysis of Industrial Wastewaters

K. H. Mancy and W. J. Weber, Jr.

The University of Michigan,
Ann Arbor

WILEY-INTERSCIENCE

a division of JOHN WILEY & SONS, INC.
*New York · London · Sydney · Toronto*

Reprinted from

*Treatise on Analytical Chemistry,*
edited by I. M. Kolthoff and P. J. Elving,
Part III, Volume 2, Section B, pages 413–562.

*Library of Congress Cataloging in Publication Data:*
Mancy, Khalil H.
    Analysis of industrial wastewaters.

    Reprinted from pt. 3, v. 2 of Treatise on analytical
chemistry, edited by I. M. Kolthoff and P. J. Elving.
    Bibliography: p.
    1. Sewage—Analysis. I. Weber, Walter J., 1934–
joint author. II. Title.

TD735.M35       628'.2       72–255
ISBN 0–471–56640–3 (pbk)

Printed in the United States of America

10 9 8 7 6 5 4 3 2 1

# Publisher's Note

The interest shown by users in "Analysis of Industrial Wastewaters" by K. H. Mancy and W. J. Weber since it appeared in the Kolthoff-Elving, TREATISE ON ANALYTICAL CHEMISTRY, Part III, Volume 2, has encouraged the Editors and Publishers to make available this reprint volume.

To keep the price at a reasonable level and to avoid difficulties in cross-referencing, the pagination has not been changed.

# ANALYSIS OF INDUSTRIAL WASTE-WATERS

K. H. Mancy and W. J. Weber, jr., *The University of Michigan Ann Arbor, Michigan*

## Contents

**Contents** (*continued*)

# I. INTRODUCTION

The quantity of water used annually for industrial operations in any highly developed society represents a significant part of the total water requirements of that society. Moreover, there is little reason to believe that with constant expansion and development of technology there will ever be a decrease in the relative use of water by industry. In 1960 an average of approximately 270 billion gallons of water was withdrawn daily from ground and surface water supplies in the United States; of this, industry accounted for 138 billion gallons per day, or about 51% of the total (278). In the same period, the average daily quantity of water consumed, i.e., not returned directly to ground or surface supplies, was 61 billion gallons, or about 23% of the total daily withdrawal. The major portion, 85%, of the daily consumption of water was for irrigation, while industry consumed 5%, and public and private water-use accounted for 10% of the total. These figures indicate that slightly more than 2% of the water used by industry was consumed and nearly 98% was returned to surface or underground sources.

Rarely is water, which has been subjected to industrial use, of the same quality as it was at the time of withdrawal from its source; rather, its quality generally has been degraded to some degree. This degradation, which may range from a simple temperature increase in the case of cooling waters to introduction of high concentrations of toxic materials in the case of some process waters, most often renders the water unfit for direct reuse in the same application, thereby necessitating its disposal as a waste. The distinction between industrial water supplies and industrial wastewaters is often not clear, for what is considered as waste for one industrial application may be suitable as a source of supply for another, and, with contiguous location of industry along a water course, multiple reuse of water is fairly common practice.

From an analytical standpoint it is desirable to strike a distinction between the two types of industrial water. In the case of industrial water supplies, chemical analyses are usually performed either for determination of the suitability of a water for use in a particular industrial process or for provision of information required for fixing the degee of treatment needed for removal of undesirable materials and/or addition of certain desirable constituents. Chemical analyses of industrial wastewater also are generally directed either to determination of the suitability of the water, in this case for reuse or disposal, or to determination of the degree of treatment required prior to disposal or for recovery of secondary products. While there seems little difference in the overall objectives of chemical analyses of industrial water supplies and wastewaters, the differences which do exist, coupled with the usually quite different nature of these two types of industrial water, often require specific test procedures of different levels of sophistication.

The present chapter deals with the chemical analysis of industrial wastewaters, with due recognition of the fact that the methodology involved is applicable, in a general sense, to all kinds of waters. An attempt has been made to present a fairly comprehensive, documented discussion of certain operating principles which are useful as guidelines for the analysis of industrial wastewaters. In this respect, the present chapter should not be considered as a substitute for any of the standard manuals on analytical procedures for waters and/or wastewaters (20,21,29,85,112, 177,212,427,428,450,456), which indeed are used as reference sources for detailed description of the more common analytical methods. The present discussion is meant to serve as a guide for the effective use of the methods described in these manuals and is intended primarily for the chemist, engineer, or other professional person concerned with all aspects of industrial wastewater analysis. Concern, therefore, is mainly with the design of measurement systems and theory of analysis rather than with stepwise procedures for analysis.

Analytical programs for industrial wastewaters must include some consideration of the major problems of water pollution associated with the discharge of wastes to streams, rivers, and other receiving waters, for these are the problems to which analysis for treatment and disposal must ultimately be related. Six major types of pollutants, any one or more of which may be associated with a particular industrial waste, are: (1) organic materials, (2) inorganic dissolved solids, (3) fertilizing elements, (4) heat, (5) suspended solids, and (6) pathogenic organisms.

Many organic pollutants undergo biochemical oxidation in receiving waters to which they are discharged, thus decreasing levels of dissolved oxygen in these waters and rendering them unsuitable for support of their natural biota and flora. Gross aesthetic damage, such as colors and odors, may also result directly from organic pollution. On the other hand, one of the most alarming aspects of organic pollution is that of the resistance of certain materials to biochemical oxidation. Such substances often escape removal by conventional treatment methods and persist for long periods, thus accumulating in receiving waters. Health and conservation agencies are presently studying the toxicity and carcinogenicity of some of these materials and their short-term and long-term effects on both man and aquatic animals and plants. The results of some recent studies have revealed, for example, that a rather definite cause and effect relationship existed between an industrial waste effluent containing pesticides and a serious fish kill in the Mississippi River (72,73). The long-term effects of such materials have not as yet been established.

Increases in the salinity of receiving waters is a water pollution problem of major concern, particularly in the southwestern parts of the United States and in most arid countries. Salinity is an expression of the total mineral content of a water, commonly expressed in per cent by weight. Increases in the salinity of fresh waters can result from such natural phenomena as intermixing with saline springs or ground waters of high salt content, from salt-water intrusion, and by evaporation from reservoirs. Certain industrial wastewaters, such as mine drainage, oil field brines, residual water from saline water conversion plants, and drainage from irrigation water, are major sources of salinity. Perhaps the greatest economic burden of saline water pollution falls on the agricultural industry, which often cannot afford the expensive process of treating very large volumes of water for salt removal.

The increased fertility of certain receiving waters, which often results in abundant blooms of algae and attached aquatic plants, can be traced to the disposal of wastewaters which are rich in phosphates and nitrates. This fertilization commonly results in the deterioration of an otherwise high quality water to the point where it may be unsuitable for domestic, industrial, or recreational uses or for the support of desirable fish and wild life.

The disposal of large volumes of high temperature wastes to receiving waters is recognized as one of the major current water pollution problems in the United States. Increases in the temperature of receiving waters may result in accelerated chemical and biological reactions, the net effects of which may be harmful to the ecological balance of streams and lakes. Prediction of the effects of thermal pollution on water quality in a given situation is difficult, requiring knowledge of such factors as physical mixing, heat exchange relationships, and biochemical reactions associated with the particular situation.

The influences of discharges of large quantities of suspended solids to receiving waters are obvious. Turbidity and excessive sediment buildup can have deleterious effects on the ecology of rivers, lakes, and streams, on the recreational value of these waters, and on their value as sources of water supply.

Advances in the technology of drinking water purification have greatly reduced the incidence of human disease resulting from bacterial pollution of receiving waters. However, outbreaks of water-borne disease still occur from time to time, usually in connection with shellfish contamination, small private water supplies, and bathing in polluted waters. There is evidence that conventional water purification processes are not completely effective in removing water-borne virus, especially if high concentrations of organic material are present. Viral and bacterial pollution usually results from the discharge of domestic, agricultural, and food processing wastewaters.

Water pollution may take many forms, but regardless of the type of pollution of concern in any instance, one may say with a fair degree of confidence that it will increase as a natural consequence of population expansion and industrial development unless definite abatement measures are taken. Industry has, for the most part, recognized the urgency of protecting our natural water resources, and, to a considerably increasing extent, is effecting treatment of wastewaters to this end. The proper control of treatment processes and the evaluation of the pollution potential of industrial wastewaters are highly dependent upon precise and accurate measurement of impurities. Thus, industrial wastewater analysis is a vital part of water pollution control.

## A. DESIGN OF MEASUREMENT SYSTEMS

The comprehensive analysis of industrial wastewaters is one of the most challenging problems with which the analytical chemist is likely to be confronted. It requires not only considerable knowledge of the application of standard analytical methods, procedures, and instrumentation, but also a keen insight into the nature of interferences and other problems, which may be quite unique to a particular waste and which may in many cases render analytical data misleading. The industrial-waste analyst must have the ability to properly interpret analytical results, pertinent observations, and the history of the water for design of an overall analytical program which will best serve for definition of a complex system which is often subject to rather wide variation in composition.

### 1. Objectives of Analysis

Definition of the purpose and objectives of analysis is the first step in the design of any measurement system; this includes the definition of particular problems to which solutions are sought. Some of the more common objectives of industrial wastewater analysis are as follows:

*a.* Estimation of possible detrimental effects of the waste effluent on the quality of a receiving water for subsequent downstream use.

*b.* Determination of the compliance of the wastewater with quality standards for water reuse, production control, or disposal in municipal sewers.

*c.* Evaluation of treatment requirements in view of water reuse.

*d.* Recovery of valuable by-products from the waste effluent.

The first objective directly concerns the problem of water pollution. The Federal Water Quality Act of 1965, Public Law 89-234, provides stringent rules prohibiting the discharge of waste effluents into interstate waters—directly or indirectly—which may result in deterioration of the quality of such waters to levels below established standards (470). Such standards are being set to protect present and future uses of our natural waters, based on economic, health, and aesthetic considerations. Furthermore, the Federal Water Quality Act of 1965 prohibits the use of any stream or portion thereof for the sole or principal purpose of transporting wastes. This strong legislative action on the part of the Federal Government has accompanied increased public and private demands for abatement of water pollution. Conscientious efforts on the part of industry are presently being made to correct problems of water pollution, and, as a result, comprehensive chemical analysis of waste effluents for better control is being emphasized strongly in most industrial waste treatment and disposal programs.

Natural bodies of fresh water can be classified according to intended use, including public water supply, fish or shellfish propagation, recreation, agricultural use, industrial water supply, hydroelectric power, navigation, and disposal of sewage and industrial wastes. Under the terms of the new Water Quality Act (470), industries which discharge liquid wastes directly or indirectly to receiving waters are obligated by Federal law to conform to specific water quality criteria or standards set by local and state agencies and approved by the Federal Water Pollution Control Administration (470). Water quality criteria, which have been reviewed rather extensively by McKee (298), Camp (84), and others may be prescribed by regional authorities within a given state, by the state itself, by interstate compacts, or by the Federal Government in cases involving interstate waters.

Methods for implementing and enforcing compliance to water quality standards vary from one state to another. Quality criteria may be established for receiving waters into which wastes are discharged; such criteria are termed "stream standards." A second method for controlling quality of receiving waters is that of setting quality standards on the effluent itself. Conservation and Public Health agencies in general favor the so-called "effluent standards." Conversely, industry in general prefers establishment of stream standards, since effluent standards most often do not provide for full use of the capacity of streams for assimilation of wastes.

Chemical analyses associated with control and treatment of industrial wastewaters for purposes of conforming to standards of water quality for streams, lakes, ocean outfalls, or underground aquifers are often more complicated than those required in instances where the standards are set for the waste effluent itself; indeed, the respective, methods of analysis and interpretation may be quite different in these two cases. In the former case the analysis is directed not only to characterization of the quality of the waste effluent, but also to determination of its effect on the ecosystem of the receiving water, while in the latter case the quality of the specific waste is the only matter of concern.

Quality standards on industrial waste effluents vary from one place to another and are primarily dependent on whether the effluent is disposed of into a natural body of water (i.e., river, lake, ocean) or into a municipal sewage treatment plant.

One common method for estimation of the deleterious effects a waste effluent will have on the quality of a receiving water is to treat the wastewater as a complete entity, thus avoiding analyses for particular constituents. For this type of evaluation the wastewater is first diluted to a level corresponding to that which would occur in the receiving water, and then certain gross parameters, such as taste, odor, color, and toxicity to fish, are measured, depending on the intended water use. While this procedure may give a preliminary indication of the ability of the waste to be assimilated harmlessly into the receiving water, its effectiveness for providing sufficient basis for any significant conclusion is highly doubtful. Among other things, the rate of self-purification of the receiving water is not accounted for in such a test procedure. Self-purification of streams and other receiving waters is a dynamic process in which the rate of biochemical transformation of pollutants is often much more significant than the ultimate assimilative capacity of the stream per se. For example, in evaluating the biochemical oxygen demand (BOD) of a particular wastewater, determination of the rate constant is at least as important as determination of the 5-day BOD. This method of analysis will be discussed in detail in a later section of this chapter.

The analysis of wastewaters which are to be discharged to municipal sewers is done principally for the purpose of evaluating compliance with certain effluent criteria set by the municipality. Effluent standards in this case are established for the purpose of protecting municipal waste treatment plants from operational interference which might be caused by industrial waste discharges and for protection of the sewer structure from damage. Both the municipality and the industry may carry out periodic analysis of the waste effluent for purposes of control and assessment of charges, which are usually related to the strength and volume of a particular waste. It is important to point out that wastewater which is discharged to municipal sewers becomes the responsibility of the municipality (484). The general requirements for acceptable wastewater for joint treatment with municipal wastes have been discussed in some detail by Byrd and others (83,484).

As far as in-plant operations are concerned, chemical analyses of industrial wastewaters are performed for one or more of the following purposes:

*a.* Estimation of material balances for processes to permit evaluation of unit efficiencies and to relate material losses to production operations.

*b.* Evaluation of continuing conformance to limits set for performance efficiency of certain unit processes.

*c.* Evaluation of the effectiveness of in-plant processes, modifications, and other measures taken for reduction of losses.

*d.* Determination of sources and temporal distributions of waste loads for purposes of by-product recovery or segregation of flows, relative to strength and type, for separate treatment.

*e.* Provision for immediate recognition of malfunctions, accidents, spills or other process disturbances.

*f.* Determination of the type and degree of treatment required for recovery of certain substances from waste effluents.

*g.* Evaluation of conformance to standards set for effluent quality and/or stream quality.

*h.* Provision for control of treatment and discharge of waste effluents according to present standards and/or according to variations in the conditions of the receiving water.

*i.* Provision of a current record of costs associated with discharge of waste effluents to municipal sewers when such costs are at least partially based on the chemical characteristics of the waste.

Some of the objectives of industrial wastewater analysis listed above are, of course, exploratory in nature and therefore occasional in frequency, while others are related to continuous or regular monitoring and control.

## 2. Choice of Parameters for Analysis

After definition of the objectives of analysis, the next step in the design of measurement systems is to decide on particular constituents for which analyses are to be made and what methods are to be employed. The analyst experienced in water quality characterization can often make the proper decision based on practiced intuition. In most cases, however, certain rather well defined guidelines should be followed.

Depending on the intended subsequent use of a receiving water, the parameters listed in Tables IA and IB are of significance for water

TABLE IA

Parameters for Water Quality Characterization—Domestic Water Supplies

| Quality parameter | Permissible criteria | Desirable criteria |
|---|---|---|
| Color (Co–Pt scale) | 75 units | <10 units |
| Odor | Virtually absent | Virtually absent |
| Taste | Virtually absent | Virtually absent |
| Turbidity | — | Virtually absent |
| Inorganic chemicals | | |
| pH | 6.0–8.5 | 6.0–8.5 |
| Alkalinity (CaCO$_3$ units) | 30–500 mg/liter | 30–500 mg/liter |
| Ammonia | 0.5 | <0.01 |
| Arsenic | 0.05 | Absent |
| Barium | 1.0 | — |
| Boron | 1.0 | |
| Cadmium | 0.01 | |
| Chlorides | 250 | <25 |
| Chromium (hexavalent) | 0.05 | Absent |
| Copper | 1.0 | Virtually absent |
| Dissolved oxygen | ≥4.0 | Air saturation |
| Fluorides | 0.8 to 1.7 mg/liter | 1.0 mg/liter |
| Iron (filtrable) | <0.3 | Virtually absent |
| Lead | <0.05 | Absent |
| Manganese (filtrable) | <0.05 | Absent |
| Nitrates plus nitrites (as mg/liter N) | <10 | Virtually absent |
| Phosphorus | 10–50 μg/liter | 10 μg/liter |
| Selenium | 0.01 | Absent |
| Silver | 0.05 | — |
| Sulfates | 250 | <50 |
| Total dissolved solids | 500 | <200 |
| Uranyl ion | 5 | Absent |
| Zinc | 5 | Virtually absent |
| Organic chemicals | | |
| Carbon chloroform extract (CCE) | 0.15 | <0.04 |
| Cyanides | 0.20 | Absent |
| Methylene blue active substances | 0.5 | Virtually absent |
| Pesticides: | | |
| Aldrin | 0.017 | — |
| Chlordane | 0.003 | — |
| DDT | 0.042 | — |
| Dieldrin | 0.017 | — |
| Endrin | 0.001 | — |
| Heptachlor | 0.018 | — |

(continued)

TABLE IA (continued)

| Quality parameters | Permissible criteria | Desirable criteria |
|---|---|---|
| Organic chemicals (continued) | | |
| Heptachlor expoxide | 0.018 | — |
| Lindane | 0.056 | — |
| Methoxyehlor | 0.035 | — |
| Organic phosphates plus | | |
| carbamates | 0.1 | — |
| Taxophane | 0.005 | — |
| Herbicides | | |
| 2,4,D plus 2,4,5-T, plus | | |
| 2,4,5-TP | 0.1 | — |
| Radioactivity | | |
| Gross beta | 1000 pc/liter | <100 pc/liter |
| Radium 226 | 3 pc/liter | <1 pc/liter |
| Strontium-90 | 10 pc/liter | <2 pc/liter |

quality characterization, and these should serve as guidelines for analysis of wastewater quality for purposes of treatment and control.

The choice of parameters for analysis depends primarily on the type of information sought. Certain tests are frequently used for the identification of various types of pollution associated with industrial wastewaters. Table II lists a number of tests and their significance.

### 3. Choice of Methods of Analysis

Choice of methods of analysis should be based on familiarity with the purpose of analysis and on the origin, properties, and intended future

TABLE IB
Parameters for Water Quality Characterization

A. Recreation and aesthetics

The general requirements are that surface waters should be capable of supporting life forms of aesthetic and recreational values. Hence, surface waters should be free from (a) materials that may settle to form objectionable deposits or float on the surface as debris, oil, and scum; (b) substances that may impart taste, odor, color, or turbidity; (c) toxic substances, including radionuclides, physiologically harmful to man, fish, or other aquatic plants or animals; and (d) substances which may result in promoting the growth of undesirable aquatic life.

Presently, there are no well defined water quality criteria for recreation or aesthetic purposes.

(continued)

TABLE IB    (Continued)

B. Aquatic life, fish and wildlife

1. *Turbidity.* Discharge of waste in receiving waters should not cause change in turbidity in the order of 50 Jackson units in warm water streams, 25 Jackson units in warm lakes, and 10 Jackson units in cold water streams and lakes.

2. *Color and transparency.* Optimum light requirements for photosynthesis should be at least 10% of incident light on the surface.

3. *Settleable matter.* Minor deposits of settleable matter may inhibit growth of flora and biota of water body. Such materials should not be discharged in surface waters.

4. *Floating matter.* All foreign floating matter should not be discharged in surface waters. A typical pollution problem is that of oil waste discharges which may (*a*) result in the formation of visible objectionable color film on the surface, (*b*) alter taste and odor of water, (c) coat banks and bottoms of water course, (d) taint aquatic biota, and (e) cause toxicity to fish and man.

5. *Dissolved matter.* The effect of dissolved matter on aquatic biota can be due to toxicity at relatively low concentrations or due to osmotic effects at relatively high concentrations. In general, total dissolved matter should not exceed 50 milliosmoles (the equivalent of 1500 mg/liter NaCl).

6. *pH, Alkalinity and acidity.* The pH range of 6.0–9.0 is considered desirable. Discharge of waste effluents should not lower the receiving water alkalinity to less than 20 mg/liter.

7. *Temperature.* Heat should not be added to a receiving water in excess of the amount that will raise the temperature by 3–5°F. In general, normal daily and seasonable temperature variations should be maintained.

8. *Dissolved oxygen.* It is generally required to maintain a dissolved oxygen level above 4–5 mg/liter. In cold water bodies it is recommended to maintain the dissolved oxygen above 7 mg/liter.

9. *Plant nutrients.* Organic waste effluents, such as sewage, food processing, canning and industrial wastes containing nutrients, vitamins, trace elements, and growth stimulants, should be carefully controlled. It is important not to disturb the naturally occuring ratio of nitrogen (nitrates and ammonia) to total phosphorus in the receiving water.

10. *Toxic matter.* Waste effluents containing chemicals with unknown toxicity characteristics should be tested and proven to be harmless in the concentration to be found in the receiving waters. Discharging pesticides in natural waters should be avoided if possible or kept below 1/100 of the 48-hr $TL_m$ values. Levels of ABS and LAS should not exceed 1.0 and 0.2 mg/liter, respectively, for periods of exposures exceeding 24 hr. It should be noticed that the presence of two or more toxic agents in the receiving water may exert a synergistic effect.

11. *Radionuclides.* No radionuclides should be discharged in natural waters to produce concentrations greater than those specified by the *United States Public Health Service Drinking Water Standards.*

C. Agricultural water use

1. *Total dissolved solids or "salinity."* This is the most important water quality consideration since it controls the availability of water to the plant through osmotic pressure regulating mechanisms. The effect of salinity on plant growth varies from one type to another and is dependent on environmental conditions.

*(continued)*

TABLE IB (continued)

2. *Trace elements.* Trace elements tolerance for irrigation waters may be summarized as follows:

| Element | Continuous water use, mg/liter | Short-term water use, fine texture soil, mg/liter |
|---------|--------------------------------|---------------------------------------------------|
| Aluminum | 1.0 | 20.0 |
| Arsenic | 1.0 | 10.0 |
| Beryllium | 0.5 | 1.0 |
| Boron | 0.75 | 2.0 |
| Cadmium | 0.005 | 0.05 |
| Chromium | 5.0 | 20.0 |
| Cobalt | 0.2 | 10.0 |
| Copper | 0.2 | 5.0 |
| Lead | 5.0 | 20.0 |
| Lithium | 5.0 | 5.0 |
| Manganese | 2.0 | 20.0 |
| Molybdenum | 0.005 | 0.05 |
| Nickel | 0.5 | 2.0 |
| Sellenium | 0.05 | 0.05 |
| Vanadium | 10.0 | 10.0 |
| Zinc | 5.0 | 10.0 |

3. *pH Acidity and alkalinity.* pH is not greatly significant and waters with pH values from 4.5 to 9.0 should not present problems. Highly acidic or alkaline waters can induce adverse effects on plant growth.

4. *Chlorides.* Depending upon environmental conditions, crops, and irrigation management practices, approximately 700 mg/liter chloride is permissible in irrigation waters.

5. *Temperature.* Very high, as well as very low, temperatures of irrigation waters can interfere with plant growth. Temperature tolerance is highly dependent on the type of plant and other environmental conditions.

6. *Pesticide.* A variety of herbicides, insecticides, fungicides, and rodenticides can be present in irrigation waters at concentrations which may be detrimental to crops, livestock, wildlife, and man. As far as the effect on plant growth and permissible levels are concerned, these variables are highly dependent on the type of chemical, type of plant, environmental factors, and exposure time.

7. *Suspended solids.* Suspended solids in irrigation waters may deposit on soil surface and produce a crust which inhibits water infiltration and seedling emergence. In waters used for sprinkler irrigation, colloids and suspended matter may form a film on leaf surfaces which impairs photosynthesis and defers growth.

8. *Radionuclides. United States Public Health Service Drinking Water Standards* are usually applied to irrigation waters.

TABLE II
Significance of Parametric Measurements

| Test or determination | Significance |
| --- | --- |
| Dissolved solids | Soluble salts may affect aquatic life or future use of water for domestic or agricultural purposes |
| Ammonia, nitrites, nitrates, and total organic nitrogen | Degree of stabilization (oxidation) or organic nitrogenous matter |
| Metals | Toxic pollution |
| Cyanide | Toxic pollution |
| Phenols | Toxic pollution, odor, and taste |
| Sulfides | Toxic pollution, odor |
| Sulfates | May affect corrosion of concrete, possible biochemical reduction to sulfides |
| Calcium and magnesium | Hardness |
| Synthetic detergents | Froth, toxic pollution |

use of the water under test. Some of the factors regarding informational requirements that should be considered in establishing methods of analysis are: (a) the required degree of sensitivity and accuracy; (b) the required frequency of analysis; and (c) the relative desirability of field and laboratory analysis.

Another point for consideration in selecting analytical methods concerns the collection, transportation, and storage of samples. Screening tests should be conducted for purposes of approximating required sample volumes, establishing desirable sites for and frequency of sampling, and providing a rough estimate of the waste composition and strength.

Listings of "standard" and "recommended" methods for analysis of natural waters and wastewaters are to be found in a variety of publications sponsored by several water works, pollution control, and public health agencies and organizations in this country and abroad (29,456). In addition, in several instances certain private industries have found it desirable to formalize listings of more specific methods for analysis of particular types of industrial wastewater (112,450).

While general procedures of analysis for specific waste constituents are highly useful, the industrial-wastewater analyst must be careful to guard against overreliance upon such procedures and against the possibility of being lulled into a false sense of security by results obtained from application of such procedures in instances where they may not be applicable. Indiscriminate application of general purpose methods for analysis without due consideration of specific interferences and other problems must be avoided. Standardization upon procedures should be made only after these procedures have been thoroughly evaluated in terms of particular analytical requirements. Continuing use of such standard methods without modification should then be subject to the condition that the characteristics of the waste being analyzed do not change significantly over the duration of the analytical program. Just as the skilled medical doctor will not prescribe treatment or medication until he has carefully examined the patient *in toto,* so the analytical chemist should select his approach to the analysis of an industrial wastewater only after making a careful diagnosis of the total problem. This diagnosis should include consideration of: (*a*) objectives of the analysis; (*b*) requirements of speed, frequency, accuracy, and precision of analysis; (*c*) effects of interferences; and (*d*) effects of systematic and environmental conditions on sampling and measurements.

## 4. Measurement Characteristics

It is desirable in the chemical analysis of industrial wastewaters to differentiate between intensive and extensive measurements. The distinction between intensive and extensive properties per se is clear. By definition, extensive properties are additive in the sense that the total value of the property for the whole of a system is the sum of the individual values for each of its constituent parts. Conversely, intensive properties are not additive and can be specified for any system without reference to the size of that system. In chemical systems, the total number of moles of a substance in a sample is considered an extensive property, while the chemical potential or molal free energy, $\mu$, of the substance is an intensive property. Consequently, analytical methods based on the titration of the number of moles of a given substance in a water sample are "extensive measurements" and they are distinctly different from "intensive measurements" based on the determination of the chemical potential or activity* of the same substance. Because of basic differences in measurement techniques, one may anticipate somewhat different

---

* Activity is an intensive parameter and is usually defined in terms of the relative fugacity. In this discussion, the term activity is used as a direct measure of difference between the partial molal free energies or the chemical potentials of a chosen and a reference state, i.e.,

$$\mu - \mu^\circ = RT \ln a \tag{1}$$

where $\mu^\circ$ is the standard chemical potential, $R$ is the gas constant, $T$ is the absolute temperature, and $a$ is the activity.

results from the two methods. Hence, potentiometric measurements of pH, pX, or pM, where X and M refer to anions and cations, respectively, may give results in contradiction to those obtained from titrimetric determinations of acidity, anions, or metal cations. This is common in the presence of certain interferences, which may cause the activity coefficient in the test solution to deviate from unity.

Similarly, in the case of voltammetric membrane electrode systems, such as the galvanic cell oxygen analyzer, the measured parameter is essentially an intensive factor, since the diffusion current is solely dependent on the difference in the chemical potentials of molecular oxygen across the membrane (286). Accordingly, values derived from measurements with a galvanic cell oxygen analyzer do not have to be equal to results obtained by titration methods for dissolved oxygen, such as the Winkler test. In the former case, the activity of molecular oxygen is the parameter measured (284,286); in the latter case, the total number of oxygen molecules present in the test volume is measured. For the majority of natural and wastewaters it is unlikely, for a variety of reasons, that the two kinds of measurements will give exactly the same results, although in many applications the difference may be negligible for all practical purposes.

The analyst should also be aware of the differences—and the significance of such differences—between the concentration (number of moles in a given volume) of a species and its chemical activity. Activity measurements are most significant in characterizing biochemical and physiochemical dynamics in aquatic environments. Consider, for example, the case of dissolved molecular oxygen, which may be involved in a variety of physical, chemical, and biological reactions in natural waters and in wastewaters. Oxygen transfer across the air–water interface or within the bulk of the aqueous phase is an example of a physical process. Under constant temperature and hydrodynamic conditions the rate of oxygen transport is solely dependent on the gradient in dissolved oxygen activity, and not always in the direction of diminishing concentration. Thus, under certain conditions, "uphill" diffusion, counteracting equilization of concentration, may occur. This is possible, if, for instance, a component not participating in diffusion causes a decrease in activity combined with an increase in concentration (salting-in effect) or an increase in activity combined with a decrease in concentration (salting-out effect). Accordingly, determination of the rate of oxygen transfer based solely on concentration measurements can be in error. In a similar fashion, in biological systems the availability of molecular oxygen for reaction depends primarily on the activity level, for diffusion through biological membranes is more precisely described in terms of oxygen activity than in terms of concentration.

For evaluation and prediction of rates of chemical reactions in waters and wastewaters, determination of the chemical activity of the substances involved is indeed highly significant. Whether a given reaction is diffusion controlled or activation controlled, knowledge of the activity is essential whenever the effects of environmental factors, such as pressure, ionic strength, solvent–solute or solute–solute interactions, are to be considered.

In addition, activity measurements can be used to provide a phenomological description of the dynamics of a system, in the form of appropriate proportionalities (e.g., Fourier's law, Ficks law, and the chemical reaction law). Description of complicated cross phenomena, such as the Soret effect and the Dufour effect are also possible based on such measurements.

### a. In Situ Analyses

Perhaps the most meaningful type of analytical program for water quality characterization is that involving *in situ* measurements. In general, methods for *in situ* analyses are to be preferred over analytical procedures which involve removal of the water from its natural environment, in the form of "grab" samples or in the form of a diverted stream, for subsequent analysis in the laboratory or field station. The main problems associated with the latter procedure can be related to the process of collecting samples, which in most cases cannot be done in such fashion as to give true representation of the test solution, and to the fact that the analyses are then usually performed under conditions quite different from those which exist at the sampling site. For example, changes in temperature and pressure may result in the escape of gases with consequent chemical or biological transformation of the species under test. When monitoring is the principal objective of the analytical program, the grab sample method usually does not provide sufficient data for a satisfactory degree of statistical significance and also can be relatively expensive on a cost per sample basis.

The use of *in situ* analysis for water quality characterization is not a novel concept. One of the classical methods for analysis of toxic compounds in water is based on the survival of fish in the natural environment. In a fashion similar to the 19th century practice of using canaries to monitor the presence of toxic gases in coal mines, *"Gambusia affinis"* has been used to monitor industrial waste effluents. In this procedure, the toxicity of the wastewater is continuously gauged by placing a certain number of fish in a net or screen box in the receiving stream and observing the number of dead or live fish at daily intervals. Death of the fish in 4–5 days is indicative of the presence of toxic substances (109). The fish, in this case, may be considered as biological indicators or sensors. Because of lack of specificity, techniques of this type find only limited application, although the "fish test" is still in fairly common use both in this country and abroad. Current methods of using biological indicators for water quality characterization will be discussed in more detail in a later part of this chapter.

Instrumental methods of *in situ* analysis are generally based upon the use of input transducers or sensors which, upon immersion in the test solution, transform responses to physical and/or chemical changes to a transmittable signal, most commonly an electrical signal. This signal is then transferred, over attenuating, amplifying, or transforming components, to a final display or readout device. The measuring system as a whole may be considered as being composed of a series of transducers, classified as (a) input or measuring transducers, (b) modifying transducers, and (c) output or read-out transducers.

Input transducers may be self-generating (active) devices, which produce an energy output for a single energy input, or non-self-generating (passive or impedance based) modules, which require more than one energy input in order to produce a single energy output. Examples of the former type are the glass electrode, the galvanic cell oxygen analyzer, and the thermocouple; conductivity cells, voltammetric cells, and resistance thermometers are examples of the latter type. Bollinger (66) and Stein (434) have recently discussed the classification and application of input transducers in measurement systems for environmental control.

*In situ* measuring systems may be based either on imbalance or on reference techniques. In the former case the output signal from the input transducer is measured directly (e.g., with a galvanometer). In the latter case, the output signal is compared to a known or reference quantity, the reference system output being varied until the difference between it and the output from the measurement system is zero.

Because such systems often involve intensive-type measurements, careful consideration should be given to the interpretation of the analytical results, especially when determinations of material balance or stream capacity for waste assimilation are concerned.

Much recent attention has been given to *in situ* analysis in conjunction with continuous monitoring systems. A partial list of parameters and sensors commonly determined *in situ* by continuous monitoring systems in surface waters and certain industrial waste effluents (309) is given in Table III.

TABLE III

Water Quality Parameters Commonly Measured by *in situ* Methods

| Quality parameters | Sensor system |
|---|---|
| Temperature | Thermistors; resistance thermometers; radiation pyrometers; thermocouples; pneumatic or capillary devices |
| Turbidity | Photoelectric cell |
| pH | Glass electrode |
| Dissolved oxygen | Voltammetric or galvanic membrane electrodes; dropping mercury electrode; gold electrode |
| Conductivity | Platinum electrode |
| Chlorides | Ag/AgCl electrode |
| Oxidation reduction potential | Platinum electrode |
| Alkalies, alkaline earths, fluorides, cyanides, and sulfides | Potentiometric membrane electrodes (specific ion electrodes) |

b. Instrumental and Automated Methods for Analysis

The preceding brief introduction to the use of instrumental systems for *in situ* analysis leads logically to a somewhat more detailed discussion of instrumental methods in general. During the last two decades there has been a noticeable increase in the use of instrumentation in all fields of chemical analysis. In the water and wastewater field, however, highly sophisticated analytical instruments and automatic analyzers have not had extensive application to date, probably due in large part to the relatively high costs involved. However, as the value of more rigid control of water quality becomes more evident, one may expect a gradual increase in the use of such instrumentation.

Instrumental methods can be very useful for obtaining the information about the physicochemical and biochemical properties of a water required for prescription of proper treatment and for control of quality. However, the analyst concerned with water quality characterization should avoid unnecessary use of complicated automatic instrumentation. Many analytical procedures lend themselves to simple manual operations without loss in accuracy or sensitivity. Such methods should be investigated thoroughly before any decision is made to use complex automated techniques. Use of the most simple instrumentation suitable for accomplishing the objectives of the analytical program is recommended, especially for water and waste treatment plants lacking personnel fully qualified to maintain and operate complicated instruments. Needless to say, small industrial and municipal treatment plants are frequently left unattended for several days; this is also true for water quality monitoring systems stationed on streams or on industrial waste effluents.

In general, the selection and use of instruments for analysis should be based on the following considerations: (1) a thorough understanding of the principle of operation of the sensor and its response characteristics; (2) a general understanding of the basic properties of the instrumental elements, their functions, and modes of operation; (3) the method of calibration of the system (certain instruments require only static calibration; others require dynamic calibration in order to determine the response of the system to rapidly fluctuating variables); (4) proper installation and use of instruments; and (5) periodic maintenance and calibration checks.

Sensors (input transducers) can generally be characterized according to their input, transfer, and output properties. The input characteristics of a sensor are specified by the type of input, the useful range of input for which the sensor can be used, and the effect of the sensor on the material under investigation. The transfer characteristics are defined by the sensitivity of the element, which is a function of the differential quotient of the output to the input quantities. Thus the sensitivity of the sensor is related to the smallest change in the measured variable that causes a detectable output.

For cases in which the transfer function is linear, the sensitivity will be constant over the range of the sensor response. The transfer function is also sometimes called the "gain," the "attenuation factor," or the "scaler factor."

Errors in measurement are generally caused by failure of the sensor to behave according to given transfer characteristics. Errors may, however, be more complex. So-called "scale errors" result from deviation of the output quantity from the true value by a constant value (e.g., zero displacement), or, on the other hand, the deviation may be a variable function of the input quantity. The output may, in certain cases, depend not only upon the applied input, but also upon the past history of the sensor. "Dynamic errors" can result when the variation of output with time does not follow precisely that of the input or when output depends upon such time functions as a time derivative or the frequency of the input (462).

The "speed of response" of a sensor refers to the time required for the sensor to undergo a certain per cent response to a sudden change in the measured variable; 90, 95, and 99% responses are common, and the particular one used for defining response speed should be specified. For sensors having first-order response rates, the term "time constant" in this case is the period required for the sensor output to respond to 63.2% $(100-100/e)$ of the stepwise change in input.

Often the difference between instrumental methods of analysis and automated chemical techniques is not sufficiently clear. Instrumental methods involve the use of sensor devices, but regardless of the degree of sophistication or complexity of the instrument, it is the analyst who initiates and controls the analytical event. Automated chemical devices, on the other hand, may or may not employ instrumental methods of analysis; they are quite often merely automated forms of classical wet chemical methods (147,292,318,338,339,482,506). Blaedel and Laessig (61) have discussed automation of analytical procedures in some length. Continuous and batch analyses have been compared and the design considerations such as response time and sample size have been presented. These authors have given a survey of continuous methods of analysis, including many of interest for the water quality area.

Colorimetric or spectrophotometric procedures are among those most suitable for automated operation. A typical automatic system of this type might be comprised of several basic components: a sample turntable, a proportioning pump, a mixing coil, a separator, a heating bath, a colorimeter, and a recorder or digital printout unit. Two schematic diagrams of two automatic analyzers are given in Fig. 1.

Samples placed on the turntable are pumped at constant rate through narrow tubing, consecutive samples being separated from each other by air bubbles. Appropriate reagents and diluents are added to each sample and allowed to mix in a coil. The sample is then driven to the separator compartment where certain interfering compounds are removed by means of dialyzing membranes, by filtration, or by other procedures. Following this, chromogenic agents are added to the sample, which then goes to a heating bath where color formation takes place. Finally, the solution is transferred automatically to a continuous flow colorimeter, in which transmittance at a preset wavelength is measured and recorded on a strip chart recorder. Known standards are periodically run through the system for measurement of transmittance. Concentrations are determined by comparing spectrophotometric data for the samples with those for the known standards. Accuracy and precision are therefore dependent upon effecting exactly the same treatment for the reference standards as for the samples to be measured. Under these conditions reactions are not necessarily carried to completion, but if exactly the same treatment is given the unknown and reference samples, the measurements are nonetheless valid on a comparative basis.

The literature concerning the use of automatic analyzers for analysis

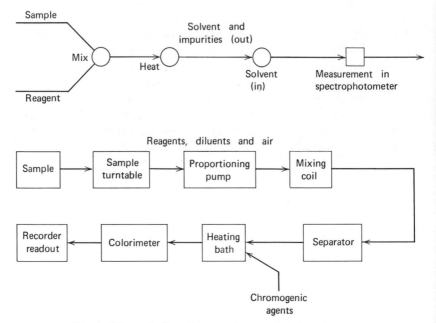

Fig. 1. Schematic flow diagrams for automated analyses.

of natural waters and wastewaters is rapidly increasing. Automatic analysis for chemical oxygen demand (COD) in domestic raw sewage and
in effluents from trickling-filter and activated-sludge processes has been
reported (318). The automated technique was compared both with the
manual COD test reported in *Standard Methods* (29) and with the
digestion method for total organic carbon reported by Weber and Morris
(487). The results of the comparisons indicated good correlation between
the manual and automatic COD procedures, but, as the authors have
pointed out, it is essential to first evaluate the suitability of a particular
automatic analyzer for each potential application and to establish correlation between the manual and automated methods. Other investigators
have reported on the use of similar automated chemical techniques for
the analysis of pesticides (506), nitrates, chlorides, and nitrogen
(252,292).

The use of automated chemical techniques for analysis of natural
waters and wastewaters should not be undertaken without careful consideration of the suitability of the underlying chemical procedure, the
physicochemical characteristics of the test solution, and the susceptibility

of the technique to pertinent interferences. If it appears feasible to employ an automatic analyzer, then the question of flexibility should be considered. It is possible to obtain equipment capable of performing more than one automatic analytical procedure. The development and operation of an automated system capable of simultaneous analysis of up to 12 different parameters has been described by Marten et al. (292).

One of the main advantages of automatic analysis is elimination of a large share of the element of human error involved in so many chemical procedures, thus enhancing reproducibility of the steps in the procedure and of the accuracy and precision of the test in general. The analyst should be aware, however, that changes in the properties of the test solution may go undetected in automated chemical procedures, leading to erroneous results. The inability of most automatic analyzers to recognize and adjust for changes in the chemical nature of test solutions is probably the principal drawback of such procedures. Because of the complex characteristics and changing nature of the test solution, it is often quite difficult, if not impossible in certain cases, to apply completely automated chemical techniques for water and waste analysis. Each sample may require a particular type of pretreatment, depending on the source, objectives of analysis, and presence of interfering substances. Because of the complexity of most automatic analyzers, required modifications of the system to this end may be difficult and costly.

#### c. Continuous Monitoring

Whenever the chemical characteristics of an industrial wastewater or receiving water are variable or whenever it is necessary to maintain close control over waste treatment processes, continuous, rather than periodic analyses are to be preferred.

The process control industry has been largely responsible for development and perfection of the equipment and "hardware" required for continuous monitoring. By far the largest application of continuous analytical measurement systems has been for industrial process control. Sophistication of continuous monitoring systems in the application has advanced to the point where not only can analyses be performed automatically and continuously, but the analytical data can be processed and translated by computer to operational control information (46). Further, this information can then be fed into automatic control devices for actual process operation.

Recently there have been substantial increases in the use of continuous monitoring systems in the wastewater treatment and quality control field (95,309,345,471). Automatic, on-line, continuous monitoring systems have been used for surveillance and remote control of wells, treatment facilities, pumping stations, storage tanks, etc. (49,105). A number of industries, including the petrochemical, pulp and paper, pharmaceutical, and a variety of chemical industries, have gone into complete or partial automation of their waste treatment facilities (46,482). Several municipal waste treatment plants also are using continuous monitoring systems for control of activated-sludge, trickling filter, and anaerobic digestion processes.

Increasing emphasis on water pollution control has led to establishment of continuous water quality monitoring stations on most of the major rivers, tributaries, lakes, and estuaries in the United States. The Public Health Service and the Geological Survey maintain a number of continuous monitoring installations, as do most interstate agencies, such as the Ohio River Valley Water Sanitation Commission (ORSANCO) and the Interstate Commission on the Delaware River Basin (471).

Industries often find it desirable to continuously monitor water quality of rivers and streams which are used as sources for water supply or for disposal of waste effluents. In the latter case, monitoring is usually done both upstream and downstream of the point of discharge for the purpose of proportioning the quality and quantity of the discharge according to the assimilative capacity of the stream and for maintenance of stream quality standards.

Continuous analytical measurements may be of the *in situ* type or may involve automated measurements made on a diverted portion of the water or waste system. In a system of the latter type the water may be pumped from the river or estuary to a small tank in which appropriate sensing elements are contained, whereas for *in situ* type systems the sensing elements are suspended directly into the river or estuary. Each of these methods has certain advantages and drawbacks. In the diversion-type stream the sensing elements are readily accessible for frequent cleaning and calibration. This convenience may, however, be gained at the expense of a truly representative monitoring system. During pumping certain changes in temperature, pressure, agitation, etc. may induce changes in the test solution which lead to absorption or release of gases or to dissolution or precipitation of salts. Other problems commonly associated with pumping the water to the monitoring station, such as clogging of the sampling lines with suspended aquatic plants, silt, etc., pump failure, and the freezing of the lines at low temperatures, may be greatly reduced by employing *in situ* techniques.

*In situ* measurements, as mentioned previously, are considered to give more significant and meaningful results, since the species under test are determined in their own environment. The main operational problems involved in *in situ* analytical systems relate primarily to protection of submerged sensors from damage by floating debris, aquatic plants, ice, etc., while still maintaining fairly easy access to the sensor equipment for purposes of frequent cleaning and checking.

It is essential for the proper design of a continuous monitoring system that the objectives of the analytical program be clearly defined, that the parameters to be measured be carefully selected, and that a well formulated plan be established for data collection, retrieval, and utilization. Design characteristics of water quality monitoring systems have been discussed in detail by Mentink (309). There are two basic variations in design, each of which has advantages for particular applications. The first is based on electronically, or mechanically and electronically, independent parametric systems, while the second is a form of an integrated system. In the first type each parametric signal is detected, transformed, amplified, and recorded independently, while the integrated system is based on the use of flexible module components for the measurement of several parameters. A well designed integrated system, beside providing all the advantages of parametric systems, is flexible, simple to maintain, and has a lower initial and operating cost (309). Modularity, by providing for simple plug-in type operation, obviates the need for design of complex operational networks, whether passive or active. Modularity in the output phase of the monitoring system extends its usefulness and application. Thus the output may not be limited to an analog strip chart recorder, but rather the addition of either telemetry or digital output equipment may be possible, this being desirable whenever large quantities of data are to be accumulated.

## II. CHARACTERIZATION OF INDUSTRIAL WASTEWATERS

The diversity of modern industry and of industrial processes makes rigid classification of wastewaters according to composition virtually impossible. Single-product industries are rare, that is, most industries process or manufacture a variety of products. Seldom is the particular combination of products processed or manufactured at a given industrial installation identical to that at another, even within the same general product field. Further, a given product is often not produced by exactly the same process at each installation at which its manufacture or preparation is carried out. Competition among industries and slight variations in grade, composition, or quality among raw materials used in production

of a given product often require continuing process innovation and modification. Variations in process type and operation usually lead to variations in the composition of the liquid waste produced. It is apparent
then that very nearly every industrial wastewater is, at least in fine
detail, unique in its composition.

Broadly, however, liquid industrial wastes may be categorized according to their principal constituents. Certain industries produce wastes
which contain primarily mineral impurities. Included in the group are
the oil field and refining, mining, steel, metal plating, foundry, and similar industries. On the other hand, liquid wastes commonly produced
by the food, beverage, and pharmaceutical industries consist primarily
of organic impurities. A number of industrial processes and operations
lead to wastewaters which contain both mineral and organic impurities
in equally significant concentrations; such wastes are common to the
paper, textile, leather, and laundry industries. Liquid waste containing
primarily radioactive impurities are associated for the most part with
water-cooled nuclear reactors and with industries involved in processing
of ores and reactor fuels. On a smaller scale as concerns both volume
and activity levels, radioactive wastes also originate in research laboratories, hospitals, and laundry operations serving such facilities. A final
category of industrial wastewater includes those of sufficiently high temperature and volume to cause thermal pollution of receiving waters into
which they are discharged. Such wastes are limited almost exclusively
to the steam power industry and to certain cooling operations within
other industries.

It must be recognized that certain complex industrial wastewaters
may fall into more than one of the above categories, but in general
the five groups listed provide a rather comprehensive classification of
liquid wastes according to their principal impurities. It should be evident,
however, even to the reader with only a casual awareness of the technology of industrial processes that marked differences must exist among
the wastes produced by the various industries which fall within any
one of the given categories.

While a broad classification of wastewaters according to the nature
of the principal constituent provides a useful sphere of reference, the
present discussion relates to analytical methods and procedures appropriate for specific industrial situations. It is appropriate then to approach
a more detailed characterization of wastewaters in terms of those typically produced by specific industries. For this purpose, six major classes
of industry are designated:

a. The chemical manufacturing industries.
b. The food, beverage, and pharmaceutical industries.

c. The apparel industries.

d. The materials industries.

e. The energy- and power-related industries.

f. The services industries.

It is entirely beyond the scope of the present discussion to consider the nature of the wastewaters associated with each of the many specific industries within the six designated classes. Rather, examples of some of the more significant types will be given. For more extensive coverage of the nature of wastes produced by specific processes and industries, the reader is advised to consult appropriate scientific periodicals, references and text books on the subject of industrial wastes. Reference works by Nemerow (332) and Gurnham (176) are examples of books in which the characterization and treatment of industrial wastewaters are discussed in detail. Such reference books also provide bibliographic listings of periodicals, reports, and other sources of information which the reader will find most useful in further exploration of this subject.

## A. CHEMICAL MANUFACTURING INDUSTRIES

There is considerable diversity among wastes produced by the various industries concerned with the manufacture and processing of chemicals. While certain similarities in particular characteristics often exist (for example, wastes originating from production of acids are of course quite low in pH while wastewaters from the manufacture of explosives, pesticides, and phosphate and phosphorus also tend to be acidic in nature), few generalities may be drawn concerning such wastes as a group.

Liquid wastes from the explosives industry are generally highly colored, quite odorous, and contain organic acids, alcohols, oils, soaps, trinitrotoluene, and dissolved metals. An average phenolphthalein acidity of about 330 mg/liter and an average pH of 2.6 have been reported for wastes from the production of TNT (423). The same report lists average concentrations of 638 and 130 mg/liter for sulfate and total nitrogen (primarily nitrate), respectively. Wastes from power plants typically contain high concentrations of mixed acids and relatively high concentrations of ether–alcohol from solvent recovery systems, as well as aniline from the manufacture of diphenylamine (332). Reports of average analytical results for wastes from ammunition production gives pH values of 3.5, a phenolphthalein acidity of 239 mg/liter, total solids of about 1700 mg/liter, suspended solids of about 400 mg/liter, and grease and copper concentrations of 709 and 86 mg/liter, respectively (423).

In addition to being acidic, pesticide wastes generally contain high concentrations of dissolved organic substances, such as benzene and other ring structures, which are quite toxic to bacteria and fish. One of the major and most troublesome constituents of wastes from the insecticide industry is dichlorophenol, resulting from the production of 2,3-dichlorophenoxyacetic acid (332). The production of DDT leads to highly acid wastes containing approximately 55% sulfuric acid, 20% ethyl hydrogen sulfate, and 20% chlorobenzene sulfuric acid (332).

The acidic wastes from production of phosphate and phosphorus typically contain slimes and tall oils. They are usually high in both suspended and dissolved solids, which consist for the most part of phosphorus, fluorides, and silica. Horton et al. (207) have described wastes from the phosphate industry as having pH values between 1.5 and 2.0 and temperatures ranging from 120 to 150°F. These investigators have reported concentrations of elemental phosphorus of between 400 and 2500 mg/liter, of total suspended solids between 1000 and 5000 mg/liter, of fluoride between 500 and 2000 mg/liter, of silica between 300 and 700 mg/liter, of $P_2O_5$ between 600 and 900 ppm, and of reducing substances between 40 and 50 mg/liter as iodine.

Several of the liquid wastes produced by the chemical manufacturing industries are characterized by high biochemical oxygen demands. Wastewaters having high BOD's include those involved in the production of cornstarch, detergents, and formaldehyde. While exhibiting high levels of BOD, formaldehyde wastes in high concentration are quite toxic to bacteria. The high BOD of cornstarch wastes is mainly attributable to starch and related materials appearing in this waste, the dry substance of the corn kernel being composed approximately of 80% carbohydrates, 10% proteins, 4.5% oil and fats, 3.5% fiber, and 2% minerals (332). Detergent wastes, which originate in the washing and purification of soaps and detergents, contain high concentrations of saponified soaps and fatty acids, as well as detergents lost in the manufacturing process.

## B. FOOD, BEVERAGE, AND PHARMACEUTICAL INDUSTRIES

Almost without exception wastewaters produced in the food, beverage, and pharmaceutical industries are high in biochemical oxygen demand, in suspended organic solids, and in dissolved organic matter. This classification of industry concerns the production and processing of canned goods, dairy products, meat and poultry products, beet sugar, yeast, pickles, coffee, fish, rice, brewed and distilled beverages, soft drinks, and pharmaceutical products.

High levels of proteins and fats are commonly found in the waste from the dairy, meat, and poultry industries, lactose additionally being high in concentration in dairy wastes and blood contributing considerably to the organic matter found in meat and poultry wastes. Process wastes from the dairy industry have been reported to have an average total solids concentration of about 4500 mg/liter, with a 5-day BOD of approximately 1900 mg/liter (332). Of the total solids, approximately 2700 mg/liter consists of organic matter and approximately 1800 mg/liter of ash solids. Approximately 4000 mg/liter of the solids are soluble, while about 500 mg/liter are suspended. These process wastes contain approximately 800 mg/liter of sodium, 112 mg/liter of calcium, 116 mg/liter of potassium, 60 mg/liter of phosphorous, 25 mg/liter of mercury, 75 mg/liter of total organic nitrogen, and about 6 mg/liter of free ammonia (332). Slaughterhouse wastes are generally quite strong, usually containing about 1000 mg/liter of total solids, 2000 mg/liter of 5-day BOD, and 500 mg/liter of total nitrogen (332). By comparison, a typical stockyard waste has been reported as having a BOD of 64 mg/liter, total suspended solids of 173 mg/liter, 8 mg/liter ammonia nitrogen, and organic nitrogen concentration of 11 mg/liter (473).

Cannery wastes are extremely variable in composition, depending upon the particular food product being processed. For example, liquid wastes from the canning of asparagus have a 5-day BOD of approximately 100 mg/liter and a suspended solids concentration of about 30 mg/liter, while wastes from the canning of pumpkin and whole kernel corn may range as high as 6000 or 7000 mg/liter in 5-day BOD, with suspended solids concentrations as high as 3000–4000 mg/liter (394). Great differences in both volume and characteristics of wastewaters from the cannery industries will be found from plant to plant, and even within the same plant from day to day. The wastewaters from all canning processes are usually made up of washwater and spillage from filling and sealing machines, which contain solids from the various sorting, peeling, and coring operations.

Beet sugar wastes are high in protein as well as sugar. Pearson and Sawyer (348) have reported that process wastewater from the beet sugar industry has a total solids concentration of about 3800 mg/liter, of which 75% is volatile and 1300 mg/liter is suspended solids. The BOD of this waste is in the neighborhood of 1600 mg/liter, while the COD is about 1500 ppm. These wastes contain about 1500 mg/liter sucrose and have a total nitrogen concentration of about 80 mg/liter, of which 65 mg/liter is protein nitrogen and 15 mg/liter is ammonia nitrogen.

Liquid wastes from the rice processing industries contain high concentrations of starch. Heukelekian has described composite rice wastes as having a BOD in the neighborhood of 1000 mg/liter with a starch level of about 1200 mg/liter and a concentration of reducing sugars of approximately 70 mg/liter (197). The total solids concentration is approximately 1500 mg/liter, of which about 11% is ash. According to Heukelekian, composite rice wastes contain about 30 mg/liter total nitrogen, and 30 mg/liter phosphates and have a pH between 4.2 and 7.0.

Fermented starches or their products are present in high concentrations in wastes from the brewed and distilled beverages industry, along with considerable concentrations of nitrogen. According to Porges and Struzeski (355) a typical waste from the carbonated beverage industry has a pH of about 10.8, with a total alkalinity of 290 mg/liter and a phenolphthalein alkalinity of 150 mg/liter. These investigators have indicated that the 5-day BOD of such wastes is about 430 mg/liter, while the suspended solids concentration is approximately 220 mg/liter.

Spent nutrient wastes from yeast plants range in total solids concentration from 10,000 to 20,000 mg/liter, with approximately 50–200 mg/liter of this being suspended solids and approximately 7000–15,000 mg/liter being volatile solids (461). This waste typically has a 5-day BOD ranging between 2000 and 15,000 mg/liter and a pH between 4.5 and 6.5, and it contains between 3800 and 5500 mg/liter total organic carbon, 500–700 mg/liter organic nitrogen, 800–900 mg/liter total nitrogen, 2000–2500 mg/liter sulfate expressed as $SO_4$, and 20–140 mg/liter phosphate expressed as $P_2O_5$.

Horton et al. (207) have reported on a coffee fermentation washwater waste having a BOD of 1700 mg/liter and a pH of 4.5, with a total solids concentration of 2100 mg/liter and a suspended solids concentration of 900 mg/liter. By comparison, coffee depulping wastes, according to Horton et al., have an average BOD of 9400 mg/liter and a pH of 4.4. The total solids concentration for the coffee depulping waste is approximately 11,500 mg/liter on the average, with approximately 800 mg/liter of this being suspended solids.

Liquid wastes from the fish processing industry can be quite odorous, while those from industries involved in the processing of pickles tend to be high in color. Pharmaceutical products can contain a broad variety of dissolved organic substances, including vitamins, trace antibiotics, and numerous other impurities, depending upon the particular product involved.

## C. APPAREL INDUSTRIES

The two principal apparel industries are those involved in the production of textiles and leather goods. Wastes originating in the textile industry from the cooking of fibers and desizing of fabrics are high in BOD, suspended solids, and temperature. In addition, such wastes tend to be high in pH and color.

Masselli et al. (294) have reported that total impurities in typical woolen mill wastes are on the order of 3800 mg/liter, of which approximately 3000 mg/liter consists of natural impurities, such as grease, suint, and dirt. Of the remaining 800 mg/liter or so of impurities contributed by process chemicals, soda ash from the scouring and fulling processes accounts for approximately 350 mg/liter, while approximately 200–250 mg/liter consists of fatty acid soaps, solvents, and detergents. These wastes also contain about 25 mg/liter of acetic acid and approximately 50 mg/liter of high carbohydrates and enzymes used in the finishing process. Other impurities to be found in lesser quantities in woolen mill wastes are pine oil, mineral oils, sulfates, chromates, phosphates, ammonia, and monochlorobenzene.

Principal waste constituents from textile mills producing synthetic fibers derive from the process chemicals used in preparing the synthetic fibers. Depending upon the particular type of synthetic fiber being processed, such wastewaters contain varying concentrations of sulfonated oils, soaps, synthetic detergents, aliphatic and fatty esters, acetic acid, ortho- and para- phenylphenol, benzoic acid, salicylic acid, phenylmethyl carbinol, formic acid, miscellaneous phenolic compounds, copper, sulfate, and ammonia (294).

Liquid wastes from the leather goods industries are also high in biochemical oxygen demand (about 1000 mg/liter), total solids (about 9000 mg/liter), pH (11–12), and total hardness (about 1600 mg/liter) (332). Typically these wastes contain about 1000 mg/liter of sulfide, approximately 1000 mg/liter of protein, 60 mg/liter ammonia nitrogen, 40 mg/liter chromium, about 300 mg/liter of sodium chloride, and about 1200 mg/liter of precipitated lime (293,332).

## D. MATERIALS INDUSTRIES

The industries included in this classification are the pulp and paper industry, the steel industry, the metal plating industry, the iron-foundry industry, the rubber industry, and the glass industry.

Wastes from the pulp and paper industries may be very high or very low in pH, depending upon the particular process used. These wastes contain inorganic fillers and are generally high in suspended colloidal and dissolved solids and in color. Pulp-mill wastes, from grinding, cooking, bleaching, and other treatment of pulps, contain sulfite, liquor, sulfides, carbonates, mercaptans, bleach, sizing, casein, ink, dyes, waxes, grease, oils, fiber, and clay (332). Paper-mill wastes, from the screens, showers, and felts of the paper machines, beaters, and mixing and regulating tanks, contain fibers, sizing, dye, and other loading material (332).

There are essentially three chemical processes for the preparation of pulps: (1) the soda process, (2) the kraft or sulfate process, and (3) the sulfite process. In place of the chemical methods either mechanical pulping or a semichemical soda process is often employed. The semichemical soda process characteristically produces wastes containing pentosans, $NaC_2H_3O_2$, formate, fine particles of bark and wood, and some dissolved solids. Moggio (317) has reported that efficiently operated kraft process plants discharge effluents containing no more than 100 lb of the sodium sulfate equivalent of the cooking liquor per ton of pulp produced. According to Moggio, the nature of the wastewater will depend to some extent on the bleaching processes employed, but, in general, suspended solids concentrations will range between 20 and 60 mg/liter (primarily fiber), with dissolved solids concentration ranging from 1000 to 1500 mg/liter. Typical kraft process wastes are highly colored and have a 5-day BOD ranging from about 100 to 200 mg/liter. Wastes from the soda pulping process may be expected to be quite similar to those from the kraft process in that the processes themselves are quite similar, the greatest difference being that in the soda pulping process either sodium hydroxide alone or a lower concentration of sodium sulfide is used in the cooking step; thus, the waste from this process will have a lower total concentration of sulfur (394). Solids concentrations of spent sulfite liquors are generally of the order of 10–12% (372), consisting mainly of lignosulfonic acid and reducing sugars (59). The sulfite liquor contains both free and combined sulfur dioxide and related sugar–sulfur dioxide derivatives, sugars, lignin, furfural, acetone, alcohols, and volatile acids (372).

Liquid wastes from the steel industry are low in pH due to the presence of considerable concentrations of mineral acids. These wastes contain a large number of dissolved and suspended materials, including limestone, oils, coke, ore, alkali, cyanogen, phenol, and mill scale. Metal-plating wastes also contain high concentrations of acid, metals, and mineral impurities, while iron-foundry wastes are high in suspended solids, mainly attributable to sand, clay, and coal.

Ammonia-still wastes from by-product coke plants represent one of the most important classes of effluent waters from steel-mill operations. Phenolic compounds represent one of the primary classes of constituents of ammonia-still wastes, with cyanides and organic and ammonia nitrogen of importance also (28). Typical BOD's for these wastes are of the order of 4000 mg/liter, with phenol levels of 2057 mg/liter, cyanide concentrations of 110 mg/liter, organic and ammonia nitrogen levels of 281 mg/liter, total suspended solids concentrations of 356 ppm, and volatile suspended solids levels of 153 ppm having been reported (28). Flue solids in wet scrubber effluents in steel mills consist primarily of iron oxide and silica, with significant percentages also of alumina, carbon, lime, and magnesia (195). Major contaminants in the wastes from the pickling and washing of steel are sulfuric acid and ferrous sulfate. Sulfuric acid concentrations are about 0.5–2.0% in pickling liquors and about 0.02–0.5% in wash waters, while corresponding concentrations of ferrous sulfate are approximately 15–22% and 0.03–0.45% respectively (332).

As far as other metal plants are concerned, the two primary wastes are those from bright-dip and pickle-bath operations. Pickle-bath wastes contain between 59.7 and 163.3 g of sulfuric acid per liter, 4.0–22.6 g of copper per liter, 4.3–41.4 g of zinc per liter, from traces to 0.56 g of chromium per liter, and from 0.1 to 0.21g of iron per liter (507). Bright-dip wastes from the brass and copper industry contain the same contaminants as do pickle bath wastes, but sulfuric acid levels are generally somewhat lower, copper levels somewhat higher, and chromium levels significantly higher, on the order of 13.5–47.7 g/liter (507).

Metal-plating wastes, generally low in volume, are extremely heterogeneous and dangerous because of their high concentrations of toxic metals. Stripping baths usually consist of mixtures of sulfuric, nitric, and hydrochloric acids and are therefore generally quite acidic in nature, although some stripping baths are alkaline, containing hydroxides, sulfides, and cyanides (332). Typical metal plating wastes have been reported as ranging in pH from as low as 2.0 to as high as 11.9, in copper concentration as high as 300 mg/liter, in iron as high as 21 mg/liter, 32 mg/liter in nickel, 82 mg/liter in zinc, 612 mg/liter total chromium, and 15 mg/liter in copper (394). Also to be found in metal plating wastes are grease, oil, and alkaline cleaners. Organic solvent cleaners are generally petroleum or coal tar emulsions, while alkaline cleaners consist of phosphates, silicates, carbonates, hydroxides, wetting agents, and emulsifiers (332).

Wastes from the production of rubber, originating from the washing of latex and coagulated rubber, generally contain high levels of suspended solids and chlorides. Rubber wastes tend to be quite variable in pH and exhibit high levels of biochemical oxygen demand and odor. Liquid wastes from the polishing and cleaning of glass have a distinctive red color, are alkaline, and contain considerable quantities of suspended solids.

Reclaimed rubber wastes have been reported with total solids concentrations as high as 63,400 mg/liter, suspended solids up to 24,000 mg/liter, BOD's up to 12,500 mg/liter, chloride concentration as high as 2000 mg/liter hydroxide alkalinities ranging to 2700 mg/liter, and pH's ranging from 10.9 to 12.2 (401). Wastes from the production of synthetic rubber are considerably less objectionable in terms of total solids and BOD. Schatz has reported synthetic rubber wastes with concentrations of total solids as high as 9600 mg/liter, suspended solids of up to 2700 mg/liter, BOD's up to 1600 mg/liter, chlorides ranging up to 3300 mg/liter, and pH ranging between 3.2 and 7.9 (401).

## E. ENERGY AND POWER-RELATED INDUSTRIES

For purposes of the present discussion this category includes those industries concerned with the processing of coals and oils, the steam power industry, and those industries concerned with nuclear power and the handling of radioactive materials.

Wastes from the coal processing industry, originating largely in the cleaning and classification of coal and in the leaching of sulfur strata with water, are high in suspended solids, due largely to the presence of coal in the liquid waste. These wastes, which include acid mine drainage wastes, are quite low in pH due to the presence of high concentrations of sulfuric acid and of ferrous sulfate. Concentrations of contaminants in acid mine drainage wastes vary considerably, with acid concentrations ranging from less than 100 mg/liter to nearly 50,000 mg/liter, sulfuric acid concentrations from 100 to 6000 mg/liter, ferrous sulfate concentrations from 10 to 1500 mg/liter, aluminum sulfate concentrations up to 350 mg/liter, and manganese sulfate from up to 250 mg/liter (332). These wastes also usually contain various concentrations of silica, sulfates, fluorides, and oxides of iron, aluminum, manganese, calcium, magnesium, and sodium (202).

Liquid wastes from the oil industry derive primarily from drilling muds, salt, oil, some natural gases, acid sludges, and miscellaneous oils from the refining process. All these wastes are high in biochemical oxygen demand, odor, and concentration of phenol. Wastes from field operations are high in dissolved salts, while those from refinery operations contain high concentrations of sulfur compounds.

A report of analyses of the mineral contents of typical oil field brines has given calcium concentrations ranging from 1507 to 12,888 mg/liter, magnesium from 346 to 4290 mg/liter, sodium from 8260 to 63,275 mg/liter, bromide from 32 to 633 mg/liter, bicarbonates from 43 to 644 mg/liter, chlorides from 12,750 to 127,220 mg/liter, sulfates as high as 1578 mg/liter, and total solids ranging from 25,210 to 248,600 mg/liter (473). Combined refinery wastes also contain crude oil and fractions thereof and dissolved and suspended mineral and organic compounds discharged in the liquors and sludges from various stages of processing (332). All refinery wastes also contain various organic sulfur compounds and organic nitrogen compounds. Typical organic sulfur compounds include mercaptans, dialkyl-sulfides, throphenes, etc. Amines, amides, quinolines, and pyridines are among the major organic nitrogen compounds found in various combinations and concentrations in oil refinery wastes.

Liquid wastes from the steam power industry, which include cooling water, boiler blowdown, and cold water drainage, are high volume wastes at elevated temperatures and contain considerable quantities of dissolved inorganic solids. Wastes other than the cooling waters which must be disposed of from steam power plants include the solutions used for cleaning boilers. These cleaning solutions, and consequently the waste discharges from the cleaning operation, contain fairly high concentrations of trisodium phosphate, sodium carbonate, sodium hydroxide, sodium sulfite, sodium nitrate, and various detergents (358). Certain waste problems can also arise from the concentration and buildup of minerals in the circulating water due to windage, leakage, and blowdown. Powell (358) has reported a concentration factor of approximately 4 for such dissolved mineral constituents as calcium, magnesium, sodium, chloride, sulfate, nitrate, bicarbonate, silica, fluoride, and boron from makeup water to circulating water.

Nuclear wastes, of course, contain radioactive elements and tend to be on the acid side. These wastes originate from the processing of ores, from the processing of nuclear fuels, from power plant cooling waters, and from research and hospital laboratory wastes.

## F. SERVICES INDUSTRIES

Two principal industries included in this classification are the laundry trades and the photographic processing industries. Laundry waste from the washing of fabrics contain considerable concentrations of soaps and detergents. These wastes are also high in turbidity, alkalinity, and organic solids. Liquid wastes from the photographic processing industries, which involves spent solutions of developer and fixer, contain various organic and inorganic reducing agents, including silver. In addition, these wastes tend to be quite alkaline.

Rudolfs (394) has reported that commercial laundry wastes have a pH of approximately 10, with a total alkalinity of about 500 mg/liter. Total solids in these wastes are in the neighborhood of 2000 mg/liter with volatile solids representing about 1500 ppm of the total. Commercial laundry wastes tend to be relatively high in grease content, at about 500 mg/liter, with a 5-day BOD in the neighborhood of 2000 mg/liter. Liquid waste from laundromats and small laundry operations, as reported from a study by Eckenfelder and Barnhart (124), generally contain between 50 and 100 mg/liter sulfonated alkylbenzenes, with a pH of between 7 and 8 and a suspended solids concentration of about 150 mg/liter. These wastes typically have a turbidity between 200 and 300 and a chemical oxygen demand between 350 and 450 mg/liter.

## III. SAMPLING

Development of a reliable system for sampling of an industrial wastewater is an important step in the overall analytical program—one which deserves the careful consideration of the analyst. The significance of a chemical analysis is no greater than that of the sampling program employed. In general, a good sampling program can be designed only by an analyst who is familiar with the physicochemical characteristics of the water to be sampled. There is no universal procedure for sampling which would be entirely suitable for all industrial wastes applications. As indicated in the preceding section, industrial wastewaters are not uniform in composition, but show appreciable changes depending on source, presence of specific or nonspecific interferences, and the effects of environmental parameters, such as temperature and pressure. There are, however, certain basic criteria or guidelines which are essential for the design of an effective sampling program. The most important requirements for a satisfactory sample are that it be both valid and representative. For a sample to be valid, it has to be one which has been collected by a process of random selection. Random selection is one of the most basic, yet most frequently violated, principles in development of a sampling program. Any method of sampling that sacrifices random selection will impair statistical evaluation of the analytical data. If nonrandom sampling procedures are contemplated, perhaps for significant reasons of convenience, it is highly desirable to first demonstrate that the results of the analysis check those which would be obtained by random sampling. This check would be essential prior to any statistical evaluation of the data.

A satisfactory sample is not only randomly drawn, but also is representative. This means that the composition of the sample should be identical to that of the water from which it was collected; the collected sample should have the same physicochemical characteristics as the sampled water at the time and site of sampling.

Planning for a sampling program should be guided by the overall objectives of analysis. Major factors of concern for any sampling program are: (a) frequency of sample collection, (b) total number of samples, (c) size of each sample, (d) sites of sample collection, (e) method of sample collection, (f) data to be collected with each sample, and (g) transportation and care of samples prior to analysis.

Frequency of sampling will depend to large extent upon the frequency of variations in composition of the water to be sampled. There are two principal types of sampling procedures commonly used for analysis of industrial wastewaters. The first type is that which yields instantaneous spot or grab samples, while the second type yields integrated continuous or composite samples. A grab sample is a discrete portion of a wastewater taken at a given time; a series of grab samples reflects variations in constituents over a period of time. The size of such individual samples will depend on the objectives and methods of analysis and on the required accuracy. The total number of grab samples should satisfy the statistical requirements of the sampling program.

Composite samples are useful for determining average conditions, which when correlated with flow can be used for computing the material balance of a stream of wastewater over a period of time. A composite sample is essentially a weighted series of grab samples, the volume of each being proportional to the rate of flow of the waste stream at the time and site of sample collection. Samples may be composited over any time period, such as 4, 8, or 24 hr, depending on the purposes of analysis.

Selection of sampling sites should be made with great care. A field survey is often useful in planning for site selection. In the case of sampling of a stream, special consideration should be given to sources of waste discharge, dilution by tributaries, and changes in surrounding topography. Sampling of streams has been discussed adequately by Velz (478) and by Haney and Schmidt (183).

Sampling of wastewater from pipes or conduits is more complicated than stream sampling, especially when the water to be sampled is under pressure. For example, in the case of a chemical treatment plant, selection of sampling sites may require extensive investigation and preliminary checking of samples from a number of effluent outlets. Proper positioning of the sampling outlet within the cross section of a conduit is essential for obtaining a representative sample, particularly for conduits of large diameter. The choice of a sampling site within the cross section of a conduit is best done by examining and comparing samples drawn from several points along the vertical and horizontal diameters of the conduit. The cross-sectional area of the opening or inlet of the sampling line should be such that the flow of water in this line is proportional to the flow of the water in the conduit. An elaborate discussion on sampling of water from pipes and conduits can be found in the ASTM *Manual on Industrial Water and Industrial Waste Water* (20).

Sampling of industrial wastewaters which contain immiscible liquids requires special consideration. Ordinary sampling techniques cannot be used in this case, since the proportionate ratio of the two immiscible liquids (e.g., mineral oil and water) usually cannot be maintained in such a sample.

The most satisfactory method of sampling two-phase or multi-phase industrial effluents is to employ a sampling tube which is capable of withdrawing a complete section of the effluent from its discharge channel or pipe. When, as is most often the case, wastewater samples are collected from the outfall of a pipe or a stream, it is suggested (427) that a large volume of the effluent be collected in a large container and left to separate, and a sectional sampling tube be used to draw the test sample.

As discussed previously, wastes discharged by industry are of great variety, and sampling must be tailored to suit the particular characteristics of a given wastewater. Sampling procedures can be expected to vary widely from one wastewater to another. Special procedures have been reported for use with waters sampled under reduced or elevated pressure and/or temperature (20). Procedures and equipment used for the sampling of waters containing dissolved gases and volatile constituents susceptible to loss upon aeration have been described by Rainwater and Thatcher (370).

A certain amount of precaution is sometimes required in sampling processes for reasons of safety. For example, strict precautionary measures (369) should be followed in taking samples from deep manholes to guard against accumulation of toxic and explosive gases and insufficiency of oxygen.

Sampling can be accomplished by either manual or automatic means, again depending on the purpose of analysis and method of sampling.

A grab sample is usually collected manually. When it is necessary to extend sampling over a considerable period of time, or when a continuous (repetitive) record of analysis at a given sampling point is required, automatic sampling equipment is commonly used.

Continuous sampling equipment, correctly designed and installed, will provide more frequent samples, tend to eliminate human errors, and in many cases be economically more feasible. A variety of automatic sampling equipment suitable for water sampling under variable conditions and for different purposes, is presently available (10,20,58,299,369,370).

The maintenance of complete records regarding the source of the sample and the conditions under which it has been collected is an inherent part of a good sampling program. This is of particular importance in field, river, or in-plant surveys, where a great number of samples are collected from different sources and under variable conditions. For illustrative purposes, the U.S. Geological Survey has defined the minimum data required for samples of surface and ground waters (370) as follows:

One of the most important aspects of the sampling process is the care and preservation of the sample prior to analysis. This point cannot be overemphasized. A water analysis is of limited value if the sample has undergone physicochemical or biochemical changes during transportation or storage. These changes are time dependent, but they usually proceed slowly. In general, the shorter the time that elapses between

| Surface Waters | Ground waters |
| --- | --- |
| Name of water body | Geographical and legal locations |
| Location of station or site | Depth of well |
| Point of collection | Diameter of well |
| Date of collection | Length of casing and position of screens |
| Time of collection | Method of collection |
| Gage height or water discharge | Point of collection |
| Temperature of the water | Water bearing formation(s) |
| Name of collector | Water level |
| Weather and other natural or other man-made factors that may assist in interpreting the chemical quality | Yield of well in normal operations |
| | Water temperature |
| | Principal use of water |
| | Name of collector |
| | Date of collection |
| | Appearance at time of collection |
| | Weather or other natural or man-made factors that may assist in interpreting chemical quality |

collection of a sample and its analysis, the more reliable will be the analytical results. Certain constituents may, however, require immediate analysis at the sample site.

Certain determinations are more sensitive than others to the method of handling of water samples before analysis. Changes in temperature and pressure may result in the escape of certain gaseous constituents (e.g., $O_2$, $CO_2$, $H_2S$, $Cl_2$, $CH_4$) or the dissolution of some atmospheric gases (e.g., $O_2$). It is recommended, therefore, that determinations for gases be done in the field, or, to "fix" such materials as $O_2$, $Cl_2$, and $H_2S$, the sample should be treated upon collection with stable oxidizing or reducing agents (284). It is also recommended that the temperature and pH of the water be determined at the site of sampling. Changes in temperature and pH may cause changes in the solubility of dissolved gases and certain nonvolatile constituents, resulting in their separation from aqueous phase. Carbonic acid–bicarbonate–carbonate equilibria may be shifted to release gaseous $CO_2$ or to precipitate certain metal carbonates. Similarly, shifts in hydrogen sulfide–sulfide equilibria due to changes in pH and/or temperature may result in the escape of $H_2S$ or the precipitation of metal sulfides.

Heavy metals ions may undergo a variety of physicochemical transformations during sample handling. It has been recommended that for analyses for Al, Cr, Cu, Fe, Mn, and Zn, samples should be filtered at the site of collection and acidified to about pH 3.5 with glacial acetic acid (370). Acidification tends to minimize precipitation, as well as sorption on the wall of the sample container. Since acetic acid may stimulate growth of molds, it may be necessary to add a small quantity of formaldehyde to the sample solution as a preservative.

Another major point of interest for handling water samples is the effect of biological activity on the sample characteristics. Microbiological activity may be effective in changing the nitrate–nitrite–ammonia balance, in reducing sulfate to sulfide, and in decreasing the dissolved oxygen content, BOD, organophosphorous compounds, and any readily degraded organic compound. Freezing of water samples is helpful in minimizing changes due to biological activity. Certain chemical preservatives, such as chlorfrom or formaldehyde, are sometimes added to water samples for this purpose (370).

That it is practically impossible to handle and process a water sample without changing its characteristics should be recognized. The best chance for error-free procedure lies in the use of *in situ* analyses. In the end, the dependability of even a well planned sampling program rests upon the experience and good judgment of the analyst.

## IV. ANALYSIS FOR MAJOR PHYSICAL CHARACTERISTICS

Methods for the analysis of industrial wastewaters have been variously classified according to: (a) the purpose of the analysis (e.g., determination of toxicity or biodegradability), (b) the nature of the constituent under test (e.g., gases, alkali metals, and heavy metals), and (c) the nature of the analytical procedure itself (e.g., titrimetric, gravimetric, and electrometric). In the present discussion, methods of analysis are somewhat arbitrarily classified according to the analytical information desired and whether it pertains primarily to the physical, chemical, or biological characteristics of the test solution. The discussion includes a survey of new methods and techniques, as well as a review of important classical analytical procedures for characterization of industrial wastewaters.

The significant physical properties of a given industrial wastewater might include: (a) density and viscosity; (b) temperature; (c) electrical conductivity; (d) turbidity; (e) particulate, volatile, and dissolved matter; (f) oils, grease, and other immiscible liquids; (g) color; (h) odor; and (i) radioactivity.

### A. DENSITY AND VISCOSITY

Literature concerning determinations of density, viscosity, and surface tension for industrial wastewaters is sparce. This probably results from the fact that the last of these tests is not frequently performed and the procedures for the first two are rather straightforward.

Differences between absolute density and specific gravity and the effect of temperature and salinity on these parameters have been discussed by Cox (99). Density determinations are of particular interest for characterization of brine wastes, for which concentrations of dissolved materials are commonly indicated by density measurements.

Density measurements of high salinity wastewaters are also significant for relating concentration on a weight basis to concentration on a volume basis. By definition, concentration expressed in terms of parts per million represents the weight of dissolved matter per one million equal weights of solution (i.e., milligrams of solute per kilogram of solution). A concentration of 1000 ppm will increase the density of solution by only approximately 0.1%, which may not be significant. At high concentrations, however, corrections must often be made to account for changes in density. The U.S. Geological Survey (370) has arbitrarily selected a concentration level of 7000 ppm, below which corrections for changes in density are not necessary.

Density determinations may be accomplished by measuring the weight of an exact volume of solution at a given temperature, commonly 20°C (the same temperature at which volumetric glassware is calibrated). Results are accurate to 10.0005 g (370). More commonly, density is measured with a hydrometer, at the temperature of the sample with appropriate correction to 20°C. Nomographs are usually provided to facilitate conversion.

Viscosity is a direct measure of the resistance of the liquid to flow or fluidity, which is of interest for certain industrial wastewaters of high solid content, waste slurries, and sludges. Results are usually expressed in centipoise units at 20°C. The viscosity of pure water is taken to be approximately equal to unity 20°C (1.009 cp).

Several instruments are available for viscosity determinations, e.g., the canal viscometer (Ostwald) or the couette viscometer (Brookfield). Viscosity can also be measured by determining the time of fall of a spherical ball of known weight and dimension through a column of the test solution (212,370).

## B. TEMPERATURE

Temperature is an intensive measure and should not be confused with the extensive property of heat capacity. Measurement of temperature in industrial waste effluents is of particular importance in cases where biochemical activity in the receiving water or the heat budget of the stream are matters of concern. Such is the case with waste cooling waters from the power industry, which may cause significant thermal pollution of receiving streams.

Liquid-in-glass thermometers, in which mercury is often used, are the simplest temperature-measuring devices. The response time of such simple thermometers is one of the longest of common temperature-measuring devices. Needless to say, liquid-in-glass thermometers are not very suitable for continuous monitoring systems. In certain cases differential temperature measurements are more significant than absolute values, and a number of commercially available thermometers can be employed.

Other temperature-measuring devices include bimetallic thermometers, radiation pyrometers, resistance thermometers, thermistors, and thermocouples. An excellent discussion of the application of temperature transducers for environmental measurements has been presented by Bollinger (66).

## C. ELECTRICAL CONDUCTIVITY

Electrical conductivity has been conveniently used as a measure of the total concentration of ionic species in a water sample (393). Much confusion exists in the literature regarding interpretation of conductivity data and the calibration of conductivity salinometers. The electrical conductance, $L$, of a solution can be represented by the expression

$$L = K_c \sum_i^n C_i \lambda_i Z_i \qquad (2)$$

where $K_c$ is a constant, characteristic of the geometry and size of the conductance cell, $C$ the molar concentration of the individual ions in solution, $\lambda$ the equivalent ionic conductance, and $Z$ the ionic charge. Thus the electrical conductance will vary with the number, size, and charge of the ions and also with some solvent characteristics, such as viscosity. For this reason a meaningful comparison of the electrical conductance of two different types of industrial wastewaters may be difficult. Equality in electrical conductance may not mean equality in total dissolved solids. Nonetheless, conductance measurements can be used to good advantage for continuous monitoring of the strength of a given wastewater. In this case a change in conductance may be assumed to be due to a change in the number of ions rather than a change in the type of ions.

Conductance measurements are used extensively for monitoring the quality of surface waters (501) and in chemical oceanography for salinity determinations (99).

Conductivity determinations are usually based on alternating current measurements using either electrodes or inductive systems, depending on how the current is generated in solution. In the more conventional system of the former type, measurement is based on the application of an alternating current (or ac potential difference) across two or more electrodes immersed in the test solution. The major disadvantages of this type of system are the possibilities for polarization and poisoning (fouling) of the electrodes. In systems of the second type, the electrodes do not come in contact with the test solution, but are isolated by a layer of glass or some other dielectric (insulating) material. A very high frequency current in the megacycle per second range is used. In such systems the conductivity cell represents a series combination of a capacitance (the dielectric material separating the electrode from the solution) and a parallel-connected capacitance and resistance (the solution capacitance and resistance, respectively). Although the cell response is less direct because the electrodes are not in contact with the solution, the problems of electrode fouling and polarization are eliminated. Conductivity measurements based on high frequency inductive systems are fundamentally more sound than those based on conventional electrode systems and eventually will gain wider acceptance for laboratory and field applications (99).

It should be noted that one of the basic problems in precision conductivity is temperature control. Temperature effects on ionic conductance in heterogenous solutions are quite complex. The temperature coefficient for a solution of constant ionic strength varies with temperature. The conductivity of sea water, for example, increases by 3% per degree increase in temperature at 0°C and only 2% per degree increase at 25°C. At 30°C the conductivity of a solution is about double the value at 0°C. Likewise, the temperature coefficient of conductivity for sea water varies appreciably with large variations in ionic strength (99,100). In oceanographic work, therefore, relative conductance is determined rather than absolute conductance. In this case the ratio of the conductance of the sample to that of a reference solution at the same temperature is measured. In certain modifications, a thermistor or a resistance is used instead of a reference solution (99).

Conductivity measurements are quite well suited for *in situ* and continuous-type analyses. Great care should be taken, however, to account for changes in temperature, pressure, and other such factors.

## D. TURBIDITY

Turbidity is a measure of the light scattering characteristics of a water and is attributable to colloidal and particulate matter suspended in the water. Reference is to a standard suspension of fine silica (20). The Jackson Candle Turbidimeter is the standard reference instrument for turbidity measurements. The Jackson turbidity test is based on measuring that length of light path through the solution at which the outline of the flame of a standard candle becomes indistinct. Results are reported in arbitrary turbidity units (48,346).

Wastes of turbidity in excess of 1000 Jackson units are diluted prior to measurement. For waters of low turbidity (less than 25 Jackson turbidity units) nephelometric or light scattering techniques are most commonly used (60,20). Black and Hannah (60) have discussed the theoretical and procedural aspects of turbidity measurements with the Jackson Candle method and more sophisticated methods. A simple low angle photometer that may be calibrated with clay suspensions in terms of Jackson turbidity units is described and recommended for use with low turbidity waters. Several commercial turbidity monitoring systems are available and have found wide use for monitoring wastewater quality (95).

## E. PARTICULATE, VOLATILE, AND DISSOLVED SOLIDS

Particulate suspended matter consists of fine, solid materials which are dispersed in water to give a heterogenous suspension. This material can be separated by filtration. Dissolved solids, on the other hand, consist of nonvolatile compounds and salts in true solution, i.e., homogeneous phase (29).

Particulate matter is usually determined by filtering a given volume of wastewater, extracting the residue with a solvent (carbon tetrachloride, benzene, or chloroform), drying at 103°C, igniting at 180°C, and weighing the final residue (29).

Dissolved solids are determined by weighing the residue from the filtrate after evaporation and extraction with organic solvent and ignition at 180°C (29).

*Standard Methods for the Examination of Water and Wastewaters* (29) classifies particulate, volatile, and dissolved matter as follows: (a) residue on evaporation, (b) total volatile and fixed residue, (c) total suspended matter, and (d) dissolved matter. Although this classification is particularly suitable for domestic wastes and municipal sewage treatment plant effluents, it is applied frequently to a variety of industrial waste effluents.

*Residue on evaporation* is determined by evaporating a given sample, drying at 103°C, and weighing the residue. *Total volatile residue* is determined by igniting the sample at 600°C after determining the residue on evaporation and calculating the weight loss due to ignition. The weight of the residue after ignition is reported as *total fixed residue*.

The *total suspended matter* is the *nonfilterable residue* and is determined by filtering a sample through a membrane filter or an asbestos mat in a Gooch crucible. The dry residue remaining after evaporation of the filtrate at 103°C is reported as *dissolved matter* of *filterable residue*. This can also be obtained by calculating the difference between the *residue on evaporation* and the *total suspended matter* (29).

Analysis for residue in an industrial wastewater generally is of little direct value in estimating its effect on a receiving water. Residue determinations are probably more valuable for control of plant operation.

## F. OILS, GREASE, AND IMMISCIBLE LIQUIDS

The gross determination of volatile and nonvolatile oily material is of particular interest for industries such as oil refineries (21). Tests for oils and grease are based on solvent extraction procedures using common solvents, such as hexane, petroleum ether, benzene, chloroform, or carbon tetrachloride. The amount of oily matter determined is primarily dependent on the type of solvent used and the extraction procedure. Needless to say, the test is not selective for immiscible oils and grease; other organic matter in solution (e.g., phenols, organosulfur compounds) will also be measured.

The procedure for determination of volatile and nonvolatile oily matter in wastewater is based on refluxing a given volume of sample and collecting the volatile oily matter, which is then measured volumetrically. The remaining sample is extracted with an immiscible solvent. The extracts are distilled to remove the solvent, and the residue is weighed and reported in units of parts per million by weight. Oily matter, measured according to this procedure, is defined (21) as hydrocarbons, hydrocarbon derivatives, and other fractions with a boiling point of 90°C or above which are extracted from water at pH 5.0 or lower using benzene as a solvent. Various extraction techniques and equipment have been reported.

Nonvolatile oily material may also be determined by flocculation of the wastewater with an iron salt, followed by extraction of the oily matter from the flocs. The sample is first acidified to pH 4 and treated with an iron salt to form a flocculent ferric hydroxide precipitate in the sample. The floc is separated from the sample by filtration and is then extracted with ether. The ether is then evaporated in a specially designed U-tube with a calibrated capillary (21). The oil is displaced into the graduated section of the tube and measured volumetrically.

An infrared spectrometric method for gross determination of volatile and nonvolatile oily matter has been described (21,420). This method is based on extraction of the oily matter with carbon tetrachloride. By means of absorption measurements of the extract at 3.4 and 3.5 $\mu$, the oily matter concentration is determined from calibration curves. This method is especially suitable for routine monitoring of effluents which are known to contain relatively constant amounts and types of oily matter (21).

## G. COLOR

It is customary to differentiate between true and apparent color in waters and wastewaters. True color is due only to matter which is in true solution, while apparent color includes the effects of matter in suspended and coloidal states as well.

The major problem associated with this aspect of the analysis of industrial wastewaters is how to define and express color. Classically, the color of a trade effluent has been determined by visual comparison with colored solutions of known concentration or with special colored glass disks. In laboratory operations, comparison is made to standard platinum–cobalt color solutions, and the standard unit of color is that produced by 1 mg of platinum per liter, in the form of chloroplatinate ion. For field use, comparison is made with colored glass disks calibrated to correspond to the platinum–cobalt scale (112).

Color determinations by visual comparison are subject to a number of interferences and variables. The main drawback to this method is the subjectivity and variation in response of different individuals to color. It is obvious also that certain industrial wastes may produce colors which cannot be matched well by the standard platinum–cobalt scale.

A more accurate determination of color in wastewaters can be accomplished by application of tristimulus colorimetry techniques (222,307,395). The color of a filtered waste can be expressed in terms which approximately describe the visual response of an individual. One of these terms relates to the brightness of color, or *luminosity*. The hue of the color (e.g., red, yellow, green) is characterized in terms of a *dominant wavelength,* and the degree of saturation (pastel, pale, etc.) is characterized by *purity*. Luminosity and purity are usually reported in units of per cent and the dominant wavelength in millimicrons.

Tristimulus parameters are commonly determined from measurements of the light transmission characteristics of a filtered sample of wastewater. Transmission data are converted to color classification terms by using standards adopted by the International Commission on Illumination (307). Chromaticity diagrams are used to describe the color numerically in terms of the tristimulus parameters (29).

Trichromatic color characteristics of filtered wastewater are measured with ordinary absorption spectrophotometers. A photometric technique has been proposed for routine work (29). This method is based on the use of three special tristimulus light filters, which, when combined with a specific light source and photoelectric cell in a filter photometer, will give effective energy distribution curves similar in shape to the "CIE" tristimulus curves (222).

## H. ODOR

Odor, like color, is a measure of a physiological response (141,165,385). Determination of odor is based solely on the olfactory senses of the analyst, or on those of a group of individuals, and on the ability of the analyst (or group) to distinguish between different levels and kinds of odors. The testing is based entirely on arbitrary comparison since no absolute units or base for odor exist (20,21).

Several authors have attempted to characterize and classify the origin of odor in wastewaters (20,39,141,387,388,432). Most of these studies treat taste and odor as closely connected human responses. Taste determinations are generally not recommended for wastewater or untreated industrial effluents and thus are excluded from the present discussion.

Odor can always be related to the presence of volatile organic and/or inorganic species present in water. Odor intensity is a function of the volatility and the concentration of the odor-causing species, as well as of certain environmental factors, such as temperature, ionic strength, and pressure. It has been claimed that there are only four basic types of odor: (a) sweet, (b) sour, (c) burnt, and (d) goaty, realizing that the many odors are in fact combinations of two or more of these groups.

Odors often can be related to the presence of certain biological forms in the wastewater, such as algae and actinomycetes. Such odor-causing organisms are believed to secrete characteristic volatile oils during growth and upon decomposition and decay. Such poetic terms as musty, earthy, woody, moldy, swampy, grassy, fishy and wet-leaves have been used to describe odors (141,387). Odors have also been classified by chemical type (20) as shown in Table IV.

Recent studies of odor characteristics and human response have led to a proposal of a steriochemical theory of odor (39,165). This theory relates the response to odor to the geometry of molecules. It has been postulated that the olfactory system is composed of receptor cells of certain different types, each representing a distinct "primary" odor, and that odorous molecules produce their effects by fitting closely into "receptor sites" on these cells. This concept is similar to the "lock and key" theory used to explain certain biochemical reactions, e.g., enzyme with substrate, antibody with antigen, and deoxyribonucleic acid with ribonucleic acid in protein synthesis.

Seven primary odors are distinguished (387), each of them by an appropriately shaped receptor at the olfactory nerve endings. The primary odors, together with reasonably familiar examples are: (a) camphoraceous, e.g., camphor or moth repellent; (b) musky, e.g., pentadencanolactone as in angelica root oil; (c) floral, e.g., phenylethyl methyl ethyl carbinol as in roses; (d) pepperminty, e.g., menthone as in mint candy; (e) pungent, e.g., formic acid or as vinegar; and (g) putrid, e.g., butyl mercaptan as in rotten eggs.

TABLE IV

Odors Classified by Chemical Types

| Odor class | Chemical types included | Odor characteristics | | | | | Odors and algae and fungi |
|---|---|---|---|---|---|---|---|
| | | Fragrance | Acidity | Burntness | Caprylicness | | |
| Estery | Esters; lower ketones | High | Medium | Low to medium | Medium | | — |
| Alcoholic | Phenols and cresols; alcohols; hydrocarbons | High | Medium to high | Low to high | Medium | | *Asterionella* *Coelosphaerium* |
| Carbonyl | Aldehydes; higher ketones | Medium | Medium | Low to medium | Medium | | *Mallemonas* |
| Acidic | Acid anhydrides; organic acids; sulfur dioxide | Medium | Very high | Low to medium | Medium | | *Anabaena* |
| Halide | Quinones; oxides (including ozone); halides; nitrogen compounds | High | Medium to high | Medium to high | Low to high | | *Dinobryon* *Actinomycetes* |
| Sulfury | Selenium compounds; arsenicals; mercaptans; sulfides and hydrogen sulfide | Medium | Medium | Very high | Very high | | *Aphanizomenon* |
| Unsaturated | Acetylene derivatives; butadiene; isoprene; vinyl monomers | High | Medium | Medium | High | | *Synura* |
| Basic | Higher amines; alkaloids; ammonia and lower amines | High | Medium | Low to medium | High | | *Uroglenopsis* *Dinobryon* |

461

It has been claimed that every known odor can be made by mixing the seven primary odors in certain combinations and proportions (388).

Odors resulting from mixtures of two or more odoriferous substances are extremely complex. The mixture may produce an odor of greater or lesser intensity than might be expected from summing the individual odors, or a completely different kind of odor may be produced (39,387,388). Accordingly, it is frequently necessary to characterize the odor of the wastewater and that of the receiving stream both separately and in combination if the actual relationship and effect are to be determined.

Odor intensity is expressed in terms of the *threshold odor number* (20,21,29). By definition, the threshold odor number is the greatest dilution of the sample that still leaves a perceptible residual odor. The test procedure is based on successive dilution of a sample with odor-free water, disregarding any suspended matter or immiscible substances, until a dilution is obtained which has a barely perceptible odor. It has been recommended that odor tests be run at 25 and 60°C (29) or 40 and 60°C (20). In all cases the sampling and test temperature should be reported, since the threshold odor is a function of temperature. A given sample, under fixed conditions, will emit a characteristic odor stimulus, but the response to this stimulus and the judgment based upon this response are purely subjective matters, and their interpretation may vary considerably from individual to individual (39,388,432). Consequently, it is desirable to use a panel or group of judges rather than a single analyst for both qualitative and quantitative evaluation of odors in water or wastewater samples (20).

## I. RADIOACTIVITY

Developments in the useful application of nuclear energy and its by-products have focused increasing attention on problems connected with water pollution by radioactive materials. Liquid wastes from the operation of nuclear reactors, wastes from the use of radioisotopes, and fallout from the detonation of nuclear weapons already have added measurable quantities of radioactivity, in excess of the natural or background level, to some natural waters.

Wastewaters from nuclear energy industries commonly contain materials which are in a state of nuclear instability and emit ionizing radiation. The health aspects of ionizing radiation in man's drinking water, in the air he breathes, and in the plants and animals he eats make the analysis and treatment of radioactive wastewaters of paramount importance.

Radioactive wastewaters are commonly classified as: (a) low level (radiation in the order of microcuries per liter); (b) intermediate level radiation in the order of millicuries per liter); and (c) high level (radioactivity measured in curies per liter) (439). There are, however, no rigid classifications for characterizing the strength of radioactive wastewaters. Oak Ridge National Laboratory defines high level wastes as those in the range of $2 \times 10^{-2}$ to $10^{-3}$ Ci/gal and low-level wastes as those having radiation levels of 10 to $10^{-1}$ $\mu$Ci/gal. On the other hand, Hanford Atomic Products Operation of General Electric Company adopts the following categorization: high level $>100$ $\mu$Ci/mg; intermediate level $10^{-5}$ to $10^2$ $\mu$Ci/ml, and low level $<10^{-5}$ $\mu$Ci/ml (128). Simple classification of radioactive wastewaters as low level, intermediate, or high level can be somewhat misleading since the radiotoxicity of specific nuclides, the quantities discharged, and certain other important relative factors are not defined by these terms.

Sources of radioactive wastewater are numerous. Operations involved in the mining and processing of radioactive ores may result in the release of liquid wastes of low level radioactivity to the environment. Wastes from the processing of irradiated reactor fuel elements are very high in radioactivity, but are usually not released directly to the environment. Rather, these are generally stored in large underground tanks.

Because of their public health significance, radioactive waste effluents must be monitored continuously, and accurate records must be maintained regarding the amount of material release. Radioactive wastes are usually separated from nonradioactive wastes prior to treatment. This is common practice in the nuclear energy industry. Analyses of radioactive wastewater are directed toward: (a) determination of the level of radiation and the type and quantity of radioactive materials present and (b) determination of the effect that the release of the waste would have upon the environment. The first objective is accomplished by analyses performed at or close to the source of release. The second objective, which is more difficult to accomplish, involves sampling and analysis of various elements of the aquatic environment to which the waste is discharged. In the latter case, the physiochemical and biological characteristics of the receiving stream must be characterized before and after discharge. Samples of the receiving water, fish, phytoplankton, zooplankton, algae, aquatic plants, bottom deposits and shellfish must be collected and analyzed for radioactive content. The results of these analyses should present an overall picture of the effect of the radioactive waste discharge on the stream. Great care should be practiced in interpreting the results of analyses performed on these different types of samples. Table V lists some of the problems to be encountered and

TABLE V

Advantages and Disadvantages of Sampling Specific Media

| Medium sampled | Advantages | Disadvantages |
|---|---|---|
| Water | Simplicity, quantitative Legal standard, human consumption | Low activity levels; varying chemical composition; fallout contamination |
| Fish | Simplicity, cumulative intake, human consumption | Low activity levels; specific differences; seasonal variation; temperature effects; varying food habits |
| Mud | Cumulative intake, concentration[a] | Qualitative; movement of bottom; sampling difficulties |
| Algae (sessile) | Cumulative intake, concentration,[a] ease of collection, fixed (nonmobile) organisms | Semispecific; nonquantitative |
| Plankton | Semicumulative, concentration[a] | Sampling difficulties; seasonal variations; mobility, contamination with silt, debris, etc.; nonquantitative |
| Vegetation (aquatic) | Ease of sampling | Little knowledge, seasonal |
| Shellfish (aquatic invertebrates) | | Highly variable depending on species, feeding habits, temperature, etc. |

[a] Signifies removal from water by surface absorption, ion exchange, precipitation, etc.

information to be derived from the sampling and analysis of various elements of the aquatic environment (439).

Sampling programs for radioactive wastewaters should follow the general procedures discussed earlier. There is, however, one additional requirement not commonly imposed on the sampling of other industrial waste effluents. This is the need to prevent loss of the radioactive material to the sample container. It is sometimes necessary, therefore, to add carrier materials or chelating agents to the sample to minimize loss by sorption on the walls of the container. This is particularly significant when the amounts of nuclide present are only of the order of $10^{-12}$ g or less (20). Glass or plastic containers present less of a problem in this regard than do those made of metal (20).

Sampling of radioactive wastewater can be done manually or automatically. Manual sampling provides more flexibility for exploratory work and for spot-check surveys. It is, however, relatively expensive, time consuming, and more subject to human errors.

Automatic sampling is commonly associated with permanent installations. It can be designed to collect water samples intermittently or continuously and in proportion to the flow of the waste effluent (31).

Sampling equipment for biological tracing material and bottom deposits has been discussed by Straub (439) and others (20).

It is often necessary to measure extremely small quantities of radioactive materials in wastewaters, in which case the problem of sensitivity becomes more significant. The sensitivity of a radioactive measurement is limited by the randomness of the disintegration process, which results in setting a lower limit to the precision of analysis. Sensitivity is also a function of sample preparation, instrument type, and test procedure.

With regard to sample preparation, two types of analyses are frequently performed on radioactive waste effluents. The first is the determination of gross alpha, beta, or gamma radioactivity. The second is analysis for specific radionuclides. Sampling and sample pretreatment differ for each type of analysis.

When sampling water for gross activity purposes, representative samples of 1–2 liters are collected in polyethylene or chemically resistant containers. It is not advisable to add preservatives to suppress biological activity, but if they are added, they should be accounted for, separately, in the analysis. The radioactivity detected in the water sample can be categorized as being associated either with insoluble solids or with soluble fractions. To analyze for both types of radioactivity, the sample is vacuum filtered through cellulose acetate paper and the solids and the filter paper are then dried, saturated with ethyl alcohol, ignited, redried, and counted. Similarly, the dissolved fraction is evaporated, dried, and counted (20,36,327). It is estimated that with this procedure an appreciable part of beta radiation and as much as 50% of alpha radiation are lost by self-absorption (36).

Analyses for specific radionuclides require chemical separation. Several techniques are applicable e.g., precipitation, complexation, ion exchange, and centrifugation (20,36,50,96,400,411).

After concentration and/or separation of the sample, gross radioactivity or specific radionuclide levels are counted. With proper calibration of instruments for a given geometry of the sample container, sample size, volume, etc. quantitative measurements can be made. As far as counting instruments are concerned, detection efficiency is governed by: (a) "geometry," or fraction of radiation emitted toward the detector;

(b) "self-absorption," or failure of part of the radiation to penetrate the sample solid; and (c) "detector efficiency," or the actual fraction of radiation entering the detector and causing a response. Background suppression is usually attained by shielding the detector with lead blocks.

Commercially available instruments are reported in several articles and manufacturers' advertisements. The end-window Geiger counter is sometimes used for detection of beta and gamma activity. The internal proportional counter may be preferred, however, since it can also detect alpha activity. Automatic sample-changer and print-read-out equipment is frequently used for routine work. "Anticoincidence" counters are used for low level beta activity (43). Direct measurements of gamma-emitting radionuclides are frequently accomplished with spectrometers equipped with crystal sodium cesium iodide. Liquid scintillation and alpha scintillation counters are sometimes used (50,96,151,324,400).

A variety of specialized instruments and monitoring equipment for the analysis of radioactivity in natural waters and wastewaters have been described (337). Automatic analyzers have been described for continuous monitoring of predominant nuclides and total beta emitters in reactor coolant water (20,31,70,163,327,334).

A certain amount of precaution is, of course, required when working with radioactive wastewaters. Safety procedures and control measures have been summarized by several authors (326,328,329,439,510).

## V. ANALYSIS FOR ORGANIC COMPOUNDS

Organic constituents of industrial wastewaters are of general concern for one or more of a number of effects they may have upon receiving waters. The most readily apparent effect is that exerted directly on aesthetic quality by certain organic pollutants as a result of foaming, formation of slicks and films, offensive odors and tastes, discoloration, etc. Of equal significance are the effects of biologically oxidizable organic wastes on the depletion of dissolved oxygen levels in receiving waters, with concomitant disruption of the natural ecology of these waters. Other organic wastes may not degrade biochemically, but may rather persist and accumulate, thus constituting insidious long-term potential health hazards. The damaging effects of organic wastes on the quality of water for particular uses are summarized in Table VI.

Certain industries characteristically discharge waste effluents which are high in organic content; these include those industries concerned with petroleum refining, coke production, food processing, pharmaceutical

TABLE VI
Characteristics of Organic Pollution

| Water use | Damage to water quality for particular use |
| --- | --- |
| Domestic[a] or industrial process supply | Taste and odor; carbon demand; chlorine demand; interference with coagulation; color; corrosion promotion; carcinogenic properties; toxicity to humans |
| Production of fish | Toxicity to fish or fish food; taint fish flesh; deoxygenation of water; promotion of filamentous organisms; sludge deposits |
| Recreation | Odor; color; floating matter; suspended matter; sludge deposits |
| Agricultural irrigation | Toxicity to plants |
| Watering of livestock | Toxicity to animals |

[a] This damage applies in cases where industrial waste is discharged to municipal sewers for combined treatment with domestic waste.

preparation, and the production of fertilizers and organic chemicals in general. The waste materials are mostly impurities which have been removed during purification of the raw materials, unmarketable or uneconomically recoverable by-products, spillage and leakage, equipment wastings, water condensates, spent extracting or absorbing solutions, etc.

Organic material may exist in the form of settleable particulate matter, in the form of colloidal matter, or in the dissolved state. Dissolved organic matter can be roughly defined as that which is not retained by a membrane filter. The present discussion is concerned principally with organic matter which is present in true solution, and, to a lesser extent, with that which exists in the form of a colloidal suspension. It should be realized that the degree of solubility of an organic pollutant is highly dependent on factors such as temperature and pressure and on the physicochemical characteristics of the aqueous phase, such as the presence of salting-in or salting-out agents.

Several approaches are commonly used for characterization of organic matter in industrial waste effluents. These can be classified into two main categories. In the first category the damaging effect (pollutional effect) of the organic waste matter on the receiving water is estimated. This may be accomplished by diluting the waste effluent with the receiving water to a level corresponding to the dilution which will result in the receiving stream and by then characterizing apropriate pollution parameters, such as odor, color, carbon demand, chlorine demand, fish-flesh tainting, persistence (resistance to biodegradation), and treatability.

The second approach is based on both qualitative and quantitative analyses for the organic compounds present in the waste effluent. Analyses for organic compounds may involve either nonspecific or specific analytical methods. In the former case, analysis is done by measurement of oxidizable organic matter, such as the biochemical oxygen demand or the chemical oxygen demand, or by determining the total organic carbon or total organic nitrogen present in a given sample. Specific analysis, on the other hand, includes identification and quantitative determination of specific species present in the test solution.

## A. NONSPECIFIC ANALYSIS

Nonspecific analytical methods are often based upon measurement of quantities of oxidizable organic material, either by biologically mediated oxidation or by strictly chemical oxidation (by wet or dry combustion procedures). The procedure most commonly employed for measurement of the susceptibility of a waste to biological oxidation is the biochemical oxygen demand (BOD) test, while the chemical oxygen demand (COD) test is widely used for measurement of concentrations of organic matter oxidizable by a dichromate reflux method. Detailed descriptions of standard procedures employed for BOD and COD measurements are readily available in the literature (20,29).

### 1. Biochemical Oxygen Demand

The biochemical oxygen demand test is essentially a bioassay method involving measurement of quantities of oxygen consumed during biological oxidation of organic waste matter under controlled conditions. Because the amount of oxygen required to convert a given quantity of biologically oxidizable organic compound to carbon dioxide and water is fixed, it is possible to interpret BOD data semiquantitatively in terms of gross concentrations of organic matter, as well as oxygen-consuming tendency.

Because the concentration of dissolved oxygen present in a stream, river, or lake is of major concern from the standpoint of water pollution control, the BOD test is useful for providing an estimation of the amount of oxygen likely to be consumed by a given amount of waste upon discharge to a receiving water. By utilizing a "seed" taken directly from the receiving water in question, actual conditions can be closely paralleled in that the measured consumption of oxygen will be that corresponding to the biological activity of the particular organisms indigenous to that receiving water. There are several distinct disadvantages to the BOD test. First, a relatively long period of time (usually 5 days) is required to obtain test results. Second, it is generally assumed that oxygen utilization by nitrification is of little consequence during the first 5 days: this is not always the case; indeed, incipient nitrification often may account for a considerable portion of the 5-day oxygen demand. Third, the BOD test is subject to interference from certain substances which exhibit toxic effects on the organisms involved in the biological breakdown of organic matter. There are certain complex organic compounds usually found in industrial wastes which are at least partially resistant to biochemical oxidation; the BOD test does not provide for measurement of the concentrations of such substances.

The standard "dilution" method for determining the BOD of a wastewater is based on the general observation that the rate of biochemical degradation of organic matter is closely proportional to the amount remaining to be oxidized. Thus the rates at which oxygen is consumed in a series of dilutions of a particular wastewater should be in proportion to the respective dilution factors, provided that all other factors are equal. It must be recognized that certain inorganic constituents of industrial wastewaters, such as ferrous iron, sulfite, and sulfide, will be oxidized by molecular oxygen. Thus one must take care to differentiate between a strictly biochemical oxygen demand and a total oxygen demand. To differentiate, an "immediate 15-min. dissolved oxygen demand" (IDOD) may be determined separately, the total oxygen demand then being the sum of the IDOD and the 5-day BOD.

The IDOD may represent either immediate oxidation of substances such as ferrous iron by molecular oxygen, or it may be attributable to oxidation by iodine produced in the acidification step of the Winkler method for determining dissolved oxygen, thus decreasing the amount of iodine that would be included in the final titration with thiosulfate.

The principal disadvantages of the dilution method for BOD include the time involved in obtaining the analytical information and the so-called "sliding effect." The sliding effect, which is evidenced when different ultimate BOD levels are obtained for different dilutions, is often due to the presence of toxic substances, lack of nutrients, or lack of proper seed.

In the "manometric" method for BOD a sample is placed in a closed system at constant temperature and agitation, and oxygen consumption is measured directly, by change in pressure at constant volume if the Warburg respirometer is used and by change in volume at constant pressure if the Sierp apparatus is used.

The advantages and limitations of both the dilution and manometric methods for BOD have been discussed in some detail. Gillman and Heukelekian (164) have reported that reproducibility is approximately the same for both methods, although values obtained by the manometric method tend to be somewhat higher than those obtained by the dilution method. According to Jaeger and Niemitz (218) the manometric method provides better reproducibility along with the advantage of continuous observation. Lee and Oswald (259) have expressed the opinion that manometric methods are useful if a complete demand curve is desired, but that they are unsuitable for large scale testing programs. Arthur (34) has stated that it is unrealistic and uneconomical to use conventional monometric methods to determine 5-day BOD, but that the advantage of a continuous record of oxygen demand as obtained by manometric methods can be achieved by using an automated respirator to determine the oxygen demand over periods shorter than 5 days. Arthur claims that the advantage of this method for automatically recording oxygen consumption over conventional manometric methods is the elimination of tedious gathering of data, and over the dilution method the advantage is the simplicity of the test and the fact that it provides a complete record of BOD, it being possible to determine the oxygen demand at any time without destroying the sample.

Hiser and Busch have presented a test for what they call the "total biological oxygen demand" (TBOD) of a wastewater (204). The TBOD test, which measures the disappearance of food from a biological system, according to Hiser and Busch, measures the total soluble organic content of a waste or substrate that is amenable to microbiological metabolism.

In the conventional BOD test, the oxygen-consuming tendency of a waste is measured in terms of the change in concentration of dissolved oxygen, as a function of time over an extended period, in a diluted sample held at constant temperature in a completely filled, closed container. The test provides a measure of the oxygen utilized for biological stabilization of the organic matter in the sample under the specific test conditions. However, its usefulness for predicting conditions in a receiving water to which the waste is to be discharged is limited, first by the relatively small, and possibly nonrepresentative, sample generally used and, second, by the absence during the test of mixing and reaeration which would normally occur in the receiving water.

The progression of the oxygen change in the BOD test has been observed to approximate a first-order reaction, disregarding any initial lag in oxygen uptake due to lack of seed or need for acclimation of the system. Streeter and Phelps (440) have described the nature of this reaction: "The rate of the biochemical oxidation of organic matter is proportional to the remaining concentration of unoxidized substance, measured in terms of oxidizability." Mathematically this may be expressed,

$$\frac{dL}{dt} = -k_1 L \tag{3}$$

which upon integration yields,

$$L_t = L_0 e^{-k_1 t} \tag{4}$$

$$Y_t = L_0(1 - e^{-k_1 t}) \tag{5}$$

where $L_0$ is the total amount of oxidizable organic matter present at initiation of the test, i.e., the ultimate BOD, $L_t$ is the amount of oxidizable organic matter remaining at time, $t$, $k_1$ is the reaction rate constant, $t$ is the elapsed time from initiation of the test, and $Y_t$ is the amount of BOD "exerted" up to time $t$.

One may certainly question—and many have—the use of such a simple expression for description of a complex biochemical process. The formulation of the BOD reaction has been widely discussed, and numerous modifications have been proposed. It is beyond the scope of the chapter to cover this aspect of the BOD test, and the reader is referred to other sources for detailed discussions of this matter (140,142,320,342,343,440,-454,457,485,509,512). Suffice it to say for the present that the expressions given in eqs. (2) and (3) are most commonly used for calculation of the standard BOD reference parameters, namely $L_0$ and $K_1$.

BOD tests are normally carried out at an incubation temperature of 20°C. The reaction is, of course, strongly dependent upon temperature, the rate of oxygen utilization increasing with increasing temperature to a maximum value beyond which further rise in temperature cannot be tolerated by the bacteria. Fair has summarized the nature of the temperature dependence of the BOD reaction (142).

Another factor which must be considered in conducting the BOD test is the factor of illumination. If the test solution contains photosynthetic bacteria or algae, care must be taken to carry out the test either in the absence of light or under uniform conditions of illumination so that the effects of these organisms are held relatively constant from one run to the next.

Toxic materials in an industrial waste can seriously affect determinations of BOD. If such materials are present one may obtain low values for BOD which are grossly in error simply because biological activity was retarded or even completely arrested by the toxic substance(s).

One way to determine whether this influence is present is to mix a quantity of the waste with a well-seeded solution of glucose, [dextrose], or some other readily oxidized organic material. If this mixture does not exert an appropriate BOD, then one must conclude that toxic materials are inhibiting the biochemical oxidation of the organic matter. If the toxic substance cannot be easily separated from the solution prior to running the BOD test (e.g., by precipitation, etc.) and if the toxic effect persists upon dilution of the waste, then the analyst should discard the BOD test in favor of a chemical method, such as the COD test.

Of considerable importance for the application of biological methods for assessment of total organic waste load is a consideration of the susceptibility of the organic constituents of a particular industrial waste to biochemical oxidation. Ludzack and Ettinger have reported on rather extensive studies of the resistance of various types of organic pollutants to biological attack (269). The results of these studies, along with reports by other investigators, are summarized below.

Most hydrocarbons on which information is available are at least moderately resistant to biological oxidation. Alkylbenzene structures with short side chains appear to exhibit fairly high resistance to oxidation. There have been reports that aromatic compounds are degraded more rapidly in natural bodies of water than are aliphatics (472). Infrared data suggest that ring opening occurs rather readily when aromatic materials are discharged to natural waters, resulting in an increase in aliphatic species (269). Davis in 1956 described tests in which cyclic hydrocarbons were found to be more resistant than aliphatic materials (110).

Most alcohols appear to be rather readily metabolized (269). Data for methanol have been found to exhibit quite a high degree of variation, with susceptibility to oxidation depending mainly on acclimation. Iso or secondary structures behave in much the same fashion as normal alcohols. Tertiary butyl and amyl alcohols and pentaerythritol are strongly resistant to biological action. Within a rather broad range, chain length does not appear to be of great importance. The diols (e.g., glycol and 1,5-pentanediol), although often reported as being highly resistant to biological oxidation, have been found by Ludzack and Ettinger to be rather readily degraded (269).

Data for only a small number of phenolic compounds can be found in the literature. Mono- and dihydric phenols or cresols seem to be quite susceptible to biological attack, and many, but not all, chlorophenols behave similarly (269). For example, 2,4,5-trichlorophenol is extremely resistant, while pentachlorophenol can be assimilated.

With the exception of benzaldehyde and 3-hydroxybutenal, the aldehydes show relatively low resistance to assimilation in acclimated systems (110). The aldehydes, however, along with the alcohols, are more resistant as a group than are the acids, salts, and esters.

Esters as a group show high resistance to biological degradation. However, if sufficient acclimation is provided, oxidation of these carbon–oxygen–carbon linkage compounds can be accomplished. Dioxane, for example, can be oxidized by specifically acclimated organisms, and diphenyl ether can also be biochemically eliminated, although it persists longer than most contaminants (110,269).

What little work that has been done with ketones indicates that they are more readily degraded than ethers, but are more resistant than alcohols, aldehydes, acids, and esters.

Amino acids are readily assimilated, with the exceptions of cystine and tyrosine. The latter structures can be broken down if sufficiently long acclimation periods are provided.

Nitrogen compounds display a wide range in metabolic availability. For amines, resistance has been noted to increase with a decrease in the number of hydrogen atoms associated with each atom of nitrogen (312). Triethanolamine, acetanilide, and purine are rather readily available to biological attack, whereas the morpholenes, with a heterocyclic ring containing both nitrogen and oxygen atoms, are highly resistant (269).

Cyanides, with the exception of iron complexes thereof, show relatively low resistance to acclimated cultures. Long acclimation times are required, however.

Nitriles are, for the most part, readily metabolized in acclimated systems. The required acclimation period varies with the specific nitrile, particularly in natural waters.

Compounds containing the vinyl group are in general degraded fairly readily, with some notable exceptions. Crotonaldehyde is unique in that at low concentrations oxidation is apparently inhibited, whereas at high concentrations it is favored (269). Side reactions may form resistant substances which are masked when an adequate excess of crotonaldehyde is present (313). Methyl vinyl ketone is one of the highly resistant exceptions referred to above. Resistance apparently increases with the formation of higher molecular weight polymers. Size and changes in reactive centers of the polymer are two contributing factors. Mills has suggested that a molecular-size borderline between resistance and availability to biochemical oxidation exists somewhere in the region of molecular weights between 250 and 600 (312).

Results from experiments on the surfactant family and analogous materials indicate that low molecular weight normal sulfonated alkylaryl compounds are not strongly resistant to biochemical oxidation, although data from Warburg and BOD studies with such materials are frequently dissimilar (269). Sulfonation appears to create a marked difference in metabolic activity of the alkylaryl compounds. Normal sulfonated butylbenzene is easily degraded, for example, while normal butylbenzene is resistant. Tertiary butylbenzene and its sulfonate derivative have both been observed to be highly resistant. This trend seems to extend through the range of 12- to 15-carbon alkyl-group compounds commonly used for the manufacture of surfactants. Restated, the normal compounds seem readily available, while tertiary compounds are not. No detergent compound tested to date has been found totally biologically inert (64,269). Sulfates, esters, and low molecular weight polyethoxy amides and esters are readily assimilated, while the higher molecular weight polyethoxy esters and amides are relatively resistant (269).

Various chlorinated compounds show wide assimilation diversity and require much more thorough study than has yet been attempted. Carbohydrates show fairly low resistance, which increases with increasing molecular weight. Lignin resistance is apparently rather high, but definitive research on this material has been impeded by the variability of the natural material and by the difficulty in obtaining unmodified lignin for purposes of study (269). Changes which occur in the material during purification generally act to increase resistance. Lignin, however, is related to coniferal alcohol or oxygenated phenylpropene structures, and it is known that both the ether and propene group contribute to resistance in other compounds (173,269).

Recently, Gates et al. (161) have pointed out the advantage of determining the rate of biochemical oxygen demand, rather than the $BOD_5$ per se. In an attempt to predict the effect of disposal of organic matter on the oxygen balance in the stream, the authors described a laboratory procedure to evaluate the assimilative capacity of receiving waters. The technique is based on the simultaneous determination of the rate of biochemical oxygen utilization and the rate of atmospheric oxygen uptake, using stream and wastewaters, in appropriate ratios.

## 2. Chemical Oxygen Demand

The standard chemical oxygen demand test measures concentrations of organic materials which may be oxidized by dichromate when refluxed for 2 hr or less in a 50% sulfuric acid medium. Depending upon the nature of the organic compounds present in a particular industrial waste, the COD test may, in the two extremes, provide either a total or a negligible measure of the organic content of the waste; more often COD data represents a fractional measure of total organic content somewhere between these two extremes.

A number of oxidants other than dichromate have been tested for possible use in the COD method, including permanganate, persulfate, ceric sulfate, perchloric acid, periodic acid, nitric acid, and numerous combinations of these and other reagents. Dichromate has been selected for the standard method on the basis of comparative testing with other reagents, such as those listed above (322). A 50% solution of sulfuric acid is employed, since it has been observed that oxidation under anhydrous conditions is more complete for the majority of compounds normally found in municipal and industrial wastewaters.

The efficiency of the dichromate reflux method for oxidation of many organic materials is enhanced considerably by the presence of a catalyst such as silver. A catalyst is required particularly for low molecular weight fatty acids and their respective salts, such as acetic acid and acetate, which are otherwise not oxidized. Since the COD method is based on measurement of dichromate consumed, ferrous iron and other oxidizable inorganic materials commonly found in industrial wastewaters present certain interferences for this test, as do chlorides. Appropriate corrections for these materials can of course be made, provided that their respective concentrations are known. In the presence of a silver catalyst, however, oxidation of chloride by dichromate is not complete, and a simple correction for chloride equivalent is not applicable. This particular interference, a very significant one for many industrial wastes, can be avoided by addition of mercuric sulfate prior to addition of the oxidant (118). The mercuric ion forms a slightly dissociated complex with the chloride and thus prevents oxidation of the latter. This method for eliminating chloride interference is applicable for chloride concentrations up to about 2000 mg/liter or slightly greater. For higher concentrations of chloride it is generally wise to reflux the sample for approximately 30 min before adding the silver catalyst. Within this period complete oxidation of chlorides is accomplished and appropriate corrections may then be applied to account for this interference. Other procedures for COD determinations in wastewaters with saline concentrations as high as 3% have been discussed by Burns and Marshall (80).

Other sources of error in the COD test include reduction of dichromate by organic nitrogen, which is in turn converted to ammonia or even nitrate in some cases, and loss of volatile organics during initial heating of the sample prior to refluxing. For application of the COD test to industrial wastes containing significant amounts of volatile organic matter, slow addition of the acid through the condenser during heating and initial stages of refluxing is recommended.

The COD test was originally intended to serve the combined functions of providing a more complete measure of the organic content of wastewaters than is afforded by the BOD test, while at the same time giving a rapid approximation to the ultimate oxygen requirements of the waste. Unfortunately, the test as presently described accomplishes neither of these objectives completely, since oxidation of many organic compounds is not complete under the conditions of the test and because certain biologically oxidizable substances are not measured as COD, while others which are not available for biological oxidation are oxidized by the dichromate–sulfuric acid mixture.

In many instances, the COD test is much more useful as a nonspecific analytical method even for estimating the oxygen requirements of industrial wastewaters than is the BOD test. It is very valuable for evaluation of wastes for which the BOD test is not applicable due to the presence of toxic materials, low rate of oxidation, or other similar factors. One can make estimates of theoretical oxygen requirements based on prior knowledge of the nature of the principal organic constituents of an industrial wastewater. For example, a sample mass balance for the reaction described in eq. (6) indicates that 192 g of molecular oxygen is required for complete oxidation of 1 mole of glucose to yield 264 g of carbon dioxide and 108 g of water. A simplified general relationship for oxidation

$$C_6H_{12}O_6 + 6O_2 \rightarrow 6CO_2 + 6H_2O \tag{6}$$

of organic matter by molecular oxygen is given by eq. (7). By comparing the percentage of the theoretical oxygen demand for a particular waste

$$C_aH_bN_cO_d + \frac{n}{2}O_2 \rightarrow aCO_2 + \frac{b}{2}H_2O + \frac{c}{2}N_2 \tag{7}$$

component obtained by the COD method with that given by the BOD test, one can evaluate the relative applicability of each of these tests for a given situation. For example, both the BOD test and the COD test (without catalyst) typically measure approximately 80% of the theoretical oxygen requirement for phenol (approximately 90% is obtained by using silver sulfate in the COD test). For a compound such as acetic acid, the percentage of the theoretical value yielded by the COD test without a catalyst is negligible, while that given by the BOD is about 80% (about 90% for the COD with silver sulfate).

As far as using the comparatively rapid COD test for estimating the 5-day biological oxygen requirement is concerned, great care must be taken not to overextend observed relationships between COD and BOD. Ratios of COD to BOD should be employed only in instances where both determinations are applicable to a given wastewater, and it should be recognized that these ratios will vary with time, with variations in the composition of the waste, with changes in prior degree of treatment of the waste, etc. If such a ratio is to be employed, and it does have certain very useful applications for predicting BOD values from measured values for COD, frequent rechecks should be made to ensure that the ratio does not change significantly over the period of concern. Normally the effect of treatment of waste on reduction of its BOD is more marked than on reduction of the COD. Thus the ratio of COD/BOD should be expected to increase with increasing degree of treatment. Dilution upon discharge of a waste to a receiving water will tend to have the same effect of increasing the COD/BOD ratio.

### 3. Total Organic Carbon

There are certain classes of perdurable organic materials occurring in increasing numbers and concentrations in industrial wastewaters which are susceptible neither to biological oxidation nor readily to chemical attack. Branched-chain synthetic surfactants, chlorinated hydrocarbons, compounds containing aromatic or heterocyclic rings, condensed ethers, and many other types of organic compounds are included in this category. Nonspecific analyses for these materials generally cannot be accomplished satisfactorily by either BOD or COD measurements. When such resistant organics are present in industrial wastewaters—indeed, even when the organics present are not particularly resistant—a most useful measure of the overall level of pollution is the total concentration of organic carbon, for carbon is, by definition, the characteristic element of organic matter.

That total organic carbon is a useful measure of pollution has been recognized for some time, and a number of techniques have been developed for this nonspecific method analysis. Without practical exception, methods for the determination of total organic carbon are based upon complete oxidation of the carbon with resultant evolution and measurement of carbon dioxide. Differences among techniques appear primarily in procedures employed for oxidation and in the particular methods used for measurement of the carbon dioxide evolved upon combustion of the organic matter.

The literature is replete with descriptions of different wet oxidation methods, and innovations thereto, in which strong oxidizing agents are used to convert organic carbon to carbon dioxide. Mohlman and Edwards (316), Van Slyke and Folch (476), Lindenbaum et al. (264), and Archer (30) have described the oxidation characteristics of various mixtures of chromic and sulfuric acids. Lindenbaum and his co-workers, one of the few groups to record data on the temperature at which the oxidation was carried out, employed digestion temperatures of 140–160°C, but, as others, failed to note the effect of temperature on the completeness of the oxidation attained. In efforts to enhance the oxidative power of the digestion medium, fuming sulfuric and orthophosphoric acids have been used in combination with chromic acid in some instances (264,476). Dichromate–sulfuric acid mixtures have received considerable attention from a number of investigators. Adeney and Dawson (11), Schulz (408), Ingols and Murray (213), and Moore et al. (321,322) have reported on studies with this medium. Ingols and Murray have suggested a digestion temperature of 145°C, reporting a rapid decrease in the oxidizing capacity of dichromate at temperatures in the neighborhood of 155°C and higher (213). Experimenting with two different concentrations—33 and 50% by volume—of sulfuric acid in the digestion mixture, Moore et al. reported better results with the 50% mixture, but did not link the increased oxidation to the fact that the more concentrated acid mixture boiled at a higher temperature (321). In this same study, Moore and his co-workers investigated the effects of adding selenium, copper, iron, nickel, and platinum to the digestion mixture, but even with these catalysts they were unable to obtain any measurable oxidation of pyridine, only about 5% recovery for acetic acid, and approximately 10% for benzene. In later work with a silver catalyst, Moore was able to obtain better results for compounds such as acetic acid (322). Description of the use of a mixture of dichromate, permanganate, and ceric sulfate has been given by Klein (237). Halate and perhalate acids have been studied by Johnson and Halvorson (221), Smith (424), and Popel and Wagner (354). Van Slyke has provided a fine summary of the historical development of wet oxidation methods (477).

Weber and Morris have described a wet oxidation method in which the oxidation is accomplished at a temperature considerably higher than temperatures normally used in other wet procedures (487). The high temperature oxidation method, effective for a large number of the types of resistant organic compounds likely to occur in industrial wastewaters, is accomplished at 175°C in a chromic acid–concentrated sulfuric acid medium.

Elimination of possible interferences from chloride in concentrations up to at least 5 g/liter is accomplished by inclusion of a saturated acidic silver arsenite trap in the combustion train, after the procedure of Pickhardt et al. (350). Interference from oxides of sulfur and nitrogen is eliminated by passing the gas stream through a solution of barium chloride in hydrochloric acid and through a tube containing silver metal and lead dioxide heated to 193°C. The method has been described as applicable over a wide range of concentrations of organic carbon and as free from interferences normally encountered in methods such as the BOD and COD. At the elevated temperature at which combustion is carried out, more complete oxidation of organic carbon is accomplished by this method than is normally achieved with other wet oxidation methods. Temperatures as high as 175°C are not practical for measurement of values of COD because reduction of the dichromate occurs with evolution of oxygen, leading to spuriously high values for COD.

A number of excellent methods for dry combustion of organic matter in the presence of a catalyst (e.g., cupric oxide, cobaltic oxide, or asbestos-supported silver permanganate at elevated temperatures of 900–1000°C) have been described, but usually these involve procedures and equipment which are relatively too sophisticated for routine analysis in the laboratories of most water and waste treatment facilities. The small samples required because of problems resulting from the evolution of large volumes of steam at the high temperatures at which combustion is carried out may be impractical from sampling and reproducibility standpoints in many instances. Furthermore, rapid combustion at very high temperature occasionally leads to incomplete oxidation to carbon monoxide rather than carbon dioxide, thus yielding spuriously low results (477). Gorbach and Ehrenberger (172), Montgomery and Thom (319), and Van Hall et al. (474) recently have described modifications of dry combustion methods. Although usually employed independently, wet and dry oxidative techniques have sometimes been used in combination (350).

There are a number of techniques for measurement of the quantity of carbon dioxide evolved upon oxidation of organic matter. One of the more elaborate means has been the use of an infrared analyzer (319,350,474). Gravimetric (504) titrimetric (94,300) conductometric (129,245), and gas chromatographic (344) methods have been employed quite commonly, and manometric procedures also have been used (476).

The carbon dioxide measured in the TOC determination cannot be correlated directly to the "oxygen demand" using the general relationship given in eq. (5) because the values for both $b$ and $d$ are unknown. Stenger and Van Hall (435) have proposed the use of carbon dioxide as an oxidizing gas in place of oxygen. Under these conditions the oxidation of organic matter may proceed as follows:

$$C_aH_bN_cO_d + mCO_2 \rightarrow (m + a)CO + \frac{b}{2} H_2O + \frac{c}{2} N_2 \qquad (8)$$

Balancing both eqs. (7) and (8) with respect to oxygen yields

$$d + n = 2a + \frac{b}{2} \qquad (9)$$

$$f + 2m = m + a + \frac{b}{2} \qquad (10)$$

Then, by subtraction and rearrangement

$$n = m + a \qquad (11)$$

From eq. (11), the number of moles of carbon monoxide produced in eq. (8) is the same as the number of oxygen atoms utilized in eq. (7). Stenger and Van Hall (435) use the above argument to show the advantage of using $CO_2$ instead of $O_2$ as an oxidizing gas to obtain a more exact stoichiometric evaluation of the "chemical oxygen demand." This procedure is best suited for oxygen demands in the range of 10–300 mg/liter.

One drawback to any method for the determination of total organic carbon is that this quantity is not a measure of the oxygen consuming capacity of an industrial waste, hence the analytical value so obtained does not directly indicate the effect of a discharge of waste on the oxygen balance of a receiving water. However, determinations of total organic carbon do offer a valuable supplement to BOD and COD determinations for estimation of the pollution potential of an industrial waste. While the concentration of organic carbon is a different type of measure of organic matter than that afforded by the BOD and the COD and, therefore, cannot be directly correlated with these other quantities, the relationship for a particular wastewater is generally close enough so that similar inferences may often be drawn and additional information obtained by comparison of determinations of organic carbon with either of the other quantities.

### 4. Total Organic Nitrogen

Organic nitrogen determinations are commonly done by the Kjeldahl method, first described some 75 years ago (243). This method is applicable to many types of organic compounds, although it is sometimes referred to as a method for aminoid or albuminoid nitrogen.

According to Kjeldahl, organic material is destroyed with sulfuric acid in the presence of various catalysts, and the nitrogen is converted to ammonium acid sulfate. The ammonia is liberated and either titrated or determined colorimetrically (20,29). One of the principal problems of this method is that certain organic-bound nitrogen structures cannot be easily transformed to ammonia (406). Four main groups can be characterized: (a) compounds with N—H linkages, (b) compounds with heterocycles, (c) compounds with N—N linkages, and (d) compounds with NO and $NO_2$ groups. The main factors which determine the degree of digestion are time, temperature, and the concentration of potassium sulfate. Mercury is considered an adequate catalyst for the digestion of group a and group b type compounds.

Compounds of groups c and d, e.g., N—N, NO, and $NO_2$, must be reduced prior to digestion. Hydroiodic acid and red phosphorous zinc (157) and other metal powders (117,276) have been used as reducing agents.

For the determination of total organic nitrogen, gaseous ammonia in solution is usually removed by distilling a buffered sample prior to the test (20,29).

Several methods utilizing various combustion techniques separately or in conjunction with the Kjeldahl procedure have been proposed. The reader is referred to the recent review articles by Schoniger (406) and Feigl et al. (146).

Automated chemical techniques are quite suitable for continuous analysis of Kjeldahl nitrogen (452). It is possible by these means to continuously monitor total nitrogen in the concentration range between 0.1 mg/liter to in excess of several thousand milligrams per liter with good correlation to classical manual methods. Appropriate procedures for homogenizing and diluting samples also have been described.

### 5. Total Organic Phosphorus

For determination of organic phosphorus, organic material is first broken down by wet combustion with mixtures of sulfuric and nitric acids. The phosphoric acid produced then can be determined either gravimetrically as ammonium phosphomolybdate or colorimetrically (20,29,449). More extensive discussion of this subject is provided in the section dealing with the anlysis of inorganic phosphates.

### B. SPECIFIC ANALYSIS

In general, specific analysis of the organic constituents of an industrial waste effluent is complicated. Direct application of conventional analytical techniques is frequently not possible because the quantities present may be well below detectable limits. Additionally, the presence of interferences may create difficulties for the application of certain separation techniques. One of the major problems associated with the isolation and separation of organic compounds from wastewaters is the biochemical degradation or chemical transformation of these compounds resulting from bacterial action, temperature effects, surface catalysis, solvent–solute and solute–solute interactions, etc.

Because of the complex nature of the majority of industrial waste effluents, specific analyses for organic compounds are not readily amenable to general regimentation and schematization. Selection of appropriate analytical and separation procedures should be based not only on the general degree of specificity and sensitivity of the test, but also on factors relating to the nature of the waste itself, factors which of course vary from one waste effluent to another.

Few analytical tests can be applied directly to an industrial wastewater without prior sample pretreatment; those which can, include the nonspecific methods discussed above together with certain methods for the analysis of phenols, proteins, and carbohydrates. In the majority of cases, however, the water sample must be subjected to one or more concentration or separation procedures prior to analysis.

## 1. Concentration Techniques

It is frequently necessary to concentrate sample solutions to bring them within the detectable limits of certain analytical procedures. Concentration by removal of water can be achieved by either evaporation or freezing. For evaporation, vacuum distillation at low temperatures is generally preferred. Partial distillation and steam distillation techniques also have been employed (308), and freeze concentration techniques have been used very effectively (41,91,416). One of the main advantages of freeze concentration is that the materials to be concentrated are kept at low temperature during the entire process, thus the chances for loss of volatile constituents and/or alteration of the nature of organic substances of interest are minimized.

Attention has recently been focused on the use of membranes for concentration of organic material in water samples (23). Reverse osmosis with cellulose acetate membranes can be used effectively to concentrate both low and high molecular weight organic compounds. In this technique, pressure greater than the osmotic pressure is applied to the sample solution, resulting in the flow of water through the membrane, with a high degree of retention of the organic solute. The advantage of the membrane-osmotic method for concentrating dilute solutions of organics is that it is an *in situ* separation technique which does not involve changes in phase or changes in temperature.

Carrying the same concept further, all the water can be removed from the sample by evaporating till dryness or freeze-drying. These techniques, however, are used primarily to get a rather rough estimate of the nonvolatile organic content.

## 2. Separation Techniques

A variety of separation techniques may be used, singularly or in combination, for isolation and concentration of organic solutes in industrial wastes. Among these are distillation (308), solvent extraction (72,391,479), precipitation and crystallization (341,353), adsorption (40), gas chromatography (41,217,491), paper chromatography (270), and thin layer chromatography (92,425).

Distillation is commonly used to separate volatile fractions. Partial distillation, in which the sample flows through the distillation cell continuously and only a small portion is distilled (10%), has been applied to oil refinery waste effluents (308). A large number of hydrocarbons can be separated from petroleum refinery effluents by a combination of distillation, selective absorption, extraction, and crystallization. Most techniques utilize fractional distillation to separate mixtures of paraffins, cycloparaffins, and aromatic hydrocarbons into fractions of roughly equal molecular weights. These groups then may be further subdivided by means of other separation techniques. Final purification may be achieved by fractional crystallization or by high efficiency distillation. Alternatively, the separate fractions from distillation may be analyzed directly, by such techniques as mass spectroscopy (308).

Precipitation techniques are useful for only a few specific applications, e.g., silver salts of acids, chloroplatinates, or tetraphenylboron derivatives of ketones and aldehydes. The precipitates may then be analyzed by X-ray diffraction, infrared spectroscopy, or other appropriate techniques (202,203). In practice, precipitation techniques find limited use because of lack of selectivity and because of the number of other substances present in the sample that may coprecipitate.

Adsorption chromatography is one of the best methods available for rapidly concentrating or extracting solutes from dilute aqueous solution. Compared with the volume of the solution used, a relatively small volume of absorbent is required for the separation. The solute normally can be recovered from the absorbent in a small volume of eluting agent.

Adsorption chromatography with activated carbon has been used for more than a decade for separating organic compounds from surface waters and certain industrial effluents (40,68,145,425). Much of the work leading to the development of this procedure was done at the Robert A. Taft Sanitary Engineering Center, Cincinnati, Ohio, in the early 1950's (72). For this procedure, a carbon column, often referred to as the "carbon filter" is used in conjunction with a sand prefilter and/or a presettling tank. The efficiency of the method is a function of the rate of flow of the sample solution, particle size of the absorbent, type and characteristics of the adsorbate(s), and physicochemical properties of the sample solution, such as temperature, pH, ionic strength, and turbidity.

The organic matter collected on the adsorption column is recovered by solvent extraction. Chloroform and ethyl alcohol have been used most commonly for extraction of organics from carbon filters (390). These two extracts are further fractionated by means of various solvent extraction schemes.

It is important to emphasize that the carbon filter technique, as it is used now (20,29,371,390), does not quantitatively separate the total organic content of a water or waste solution. Recoveries may range from approximately 50 to 90%, and replicate samples may agree within only about ±10% (29). In spite of these drawbacks, the carbon filter technique is useful for qualitative collecting of organic matter from dilute waste effluents. It can be used to advantage for screening purposes, as well as for monitoring industrial waste effluents.

Solvent extraction has been used widely for separation of fats, oils, waxes, hydrocarbons, pigments, and other organic waste products from a multitude of waste effluents. This technique can be applied in the form of batch or multiple stage processes, continuous extraction processes, or countercurrent processes. The applicability and selectivity of countercurrent extraction processes has been demonstrated by Craig and Craig (102).

Several procedures have been devised for the systematic separation of organic fractions from wastewaters by successive extraction techniques (91). Some of these techniques are associated with the carbon filter extraction procedure. In this case, the organic compounds commonly are extracted from the carbon filter with chloroform and ethyl alcohol. The weight of the residue remaining after evaporation of the solvent gives a first approximation to the total organic matter present. Further fractionation may follow Schemes 1 and 2, a procedure frequently used for separation of organic matter from river water (72).

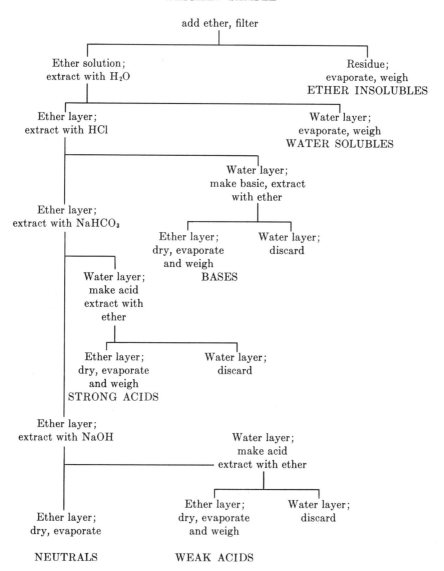

Scheme 1. Liquid extraction scheme for organic compounds.

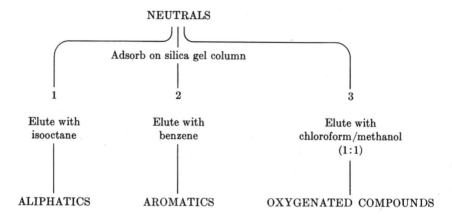

Scheme 2. Chromatographic separation of neutrals.

A countercurrent batch extraction technique has been reported for concentration of trace organic materials from large volumes of water (205). In this technique, continuous extraction is carried out with *n*-butanol for 40 hr at adjusted pH levels of 2, 7, and 10. The three extractions are then combined and condensed by vacuum distillation for further separation and identification by gas chromatography.

Gas chromatography has proven to be a useful tool for analysis of natural waters and wastewaters, being used in the majority of cases to separate and identify components of extracts from carbon filters or from other separation processes (5,25,40,175,209,330,333,341,391,483). In a few cases, however, water samples have been directly injected for analysis of organic content (38,144,491).

The successful application of gas chromatography for analysis of the organic content of a wastewater depends upon careful selection of an appropriate set of operating parameters; these parameters include column temperature, gas flow rate, type of carrier gas, gas pressure, column length and diameter, nature and particle size of the solid phase, and nature and amount of the liquid phase. Other factors requiring careful consideration include temperature programming sample injection device and type of detector. Discussions of the selection of detectors, columns, and operating conditions for certain types of waste effluents are provided in the references cites above.

Some of the factors that must be considered in selection of a detector are: sensitivity, linearity of response, frequency of calibration, response time, reproducibility, and temperature range. Both primary and secondary methods of detection have been used. Primary detectors, such as thermal conductivity and ionization detectors (sometimes referred to as nonspecific detectors), cannot specifically identify the individual organic components as they emerge from the chromatographic column. Thermal conductivity detectors are used primarily for analysis of volatile hydrocarbons, organic acids, and the components of gas streams (25). Flame ionization detectors find wider use for analysis of a variety of organic compounds by direct-injection techniques (38,491). The main advantage of the flame ionization detector for this type of analysis is that it is relatively insensitive to water and thus permits direct injection of aqueous samples.

Secondary detection involves specific analysis by such techniques as ultraviolet, infrared, and mass spectroscopy. These techniques are frequently used for the subsequent identification of organic fractions which have been separated by gas–liquid chromatography.

Paper chromatography and thin layer chromatography also find application for separation and identification of organic constituents in wastewaters. Phenols, cresols, xylenol, and other industrial and biological phenols have been determined in microgram per liter concentrations in polluted waters by paper chromatography (171). Chlorinated hydrocarbon pesticides have been determined at concentrations of 2 $\mu$g/liter or less in carbon–chloroform extracts (5).

Thin layer chromatography has been used effectively for removing interfering substances from pesticides prior to gas chromatographic analysis (425). Christman has discussed the advantages of thin layer chromatography over paper chromatography for separation and identification of organic chromogenic agents in natural waters (91). The separation and detection of various herbicides (e.g., 2,4-D; 2,4,5-T MCPA; MCPB; 2,4-DB; Dalapan, etc.) can be effectively done by thin layer chromatographic techniques (425).

Methods for continuous concentration and separation of organic material have been used in connection with automated analytical techniques and continuous monitoring systems. Continuous distillation units are suitable for removal of low boiling organic fractions from a wastewater stream (107,308). Continuous filtration and dialysis units have been used with the Technicon Autoanalyzer (451). Similarly, continuous liquid–liquid extraction (205,227,398) and continuous gas chromatographic techniques (349) have been applied for separation of low boiling organics.

### 3. Identification Techniques

One of the more common methods for identification of organic compounds is the detection of functional groups by means of color reactions. Many schemes for classification by solvation, general chemical reactions, and special tests are found in most texts on organic analysis. Chelation methods have been used recently for detection of functional groups in various compounds (430). Several comprehensive surveys dealing with this aspect of organic analysis are available in the literature (87).

Spectrometric identification of organic compounds by infrared, ultraviolet, mass spectroscopy, and nuclear magnetic resonance techniques have been reviewed in previous chapters of this Treatise. For analysis of natural water and wastewaters, analytical spectrometry must generally be preceded by separation of the organic material from the water sample by one or more of the techniques discussed previously (72,92,308,391,479,496). Special procedures are necessary for collection and concentration of the sample after separation. Liquid samples in a solvent matrix, or solid potassium bromide mulls or paste in mineral oil, may be analyzed for typical infrared absorption bonds to provide structural identification. Near-infrared spectrometry offers some advantages when interferences occur in the "fingerprint" region from 7 to $12\mu$. Measurement in the near-infrared region $(1–3\mu)$ consists of overtone of hydrogen stretching vibration. These spectra are useful only for detecting and subsequently identifying functional groups which contain unique hydrogen atoms and are not characteristic of the organic molecule as a whole. Most useful are the stretching modes of vibration of the hydrogen bond in combination with carbon, oxygen, nitrogen, or sulfur.

A modification of infrared spectrometry which offers unique possibilities for analysis of organic matter in water involves attenuated total reflectance (139,186,187). This method is based on the passage of a monochromatic or monochromatically scanned light into a crystal of suitable material, in contact with the test solution, and detection of the transmitted light intensity. Changes in light intensity are related to changes in the type and concentration of "light active" substances in contact with the crystalline material. The principle is based on energy reflection at the interface between media of different refractive indices. Little sample preparation is required, and the use of two attenuation attachments permits differential spectral attachment (291,502).

Infrared scanning has been used recently for pollution surveillance (475). This technique is particularly applicable for determination of the origin and distribution of thermal pollution. The procedure involves noncontact, infrared aerial mapping. By means of air-borne detector systems, energy in the nonvisible portion of the electromagnetic spectrum (wavelength 8–14 $\mu$) is converted, recorded, and correlated with broadband emission properties of the materials viewed. Such properties are a function of temperature and material emissivity characteristics, and a theory has been developed to separate temperature and emissivity contributions (475).

Ultraviolet spectrometry has not been widely used for analysis of natural waters and wastewaters, despite an abundance of literature on the application of uv-visible spectrometry for organic analyses. Exhaustive coverage of this subject can be found in reviews appearing in some recent periodicals (e.g., *Journal of Molecular Spectroscopy, Spectrochimica Acta, Applied Spectroscopy, Talanta,* and *Analytica Chimica Acta*).

Phenolic compounds in wastewaters may be determined by uv spectrometry by a technique based on comparing the bathochromatic shifts in wavelength in alkaline solutions and in neutral solutions (496). This method is particularly useful when organic separation is difficult.

Ultraviolet spectrometry finds wide application in industrial processes for monitoring the composition of reaction mixtures. Monitoring of trace organic material in industrial effluents by uv spectrometry has been reported by Bramer et al. (71). The method is based on measuring uv absorption spectra for compounds having aromatic or conjugated unsaturated molecular configurations. The instrument employed for these analyses uses a mercury discharge lamp with a principal uv radiation at 2537 Å. The author has reported a sensitivity of 10 mg/liter of phenol in water.

Mass spectrometry has not as yet been widely used for analysis of wastewaters. Melpolder et al. (308) have described the distillation of volatile organic compounds from oil refinery wastewater and subsequent identification by mass spectrometry. Because of the high training level required for operating personnel, the data handling problems, and the cost of instrumentation, this technique has been restricted primarily to research applications, and then only at a few major water pollution control laboratories. The technique does, however, offer considerable possibilities; for example, mass spectrometry has been used to detect phenols in nanogram quantities in the effluent from a gas chromatographic column (169).

Fluorescence spectrophotometry has significant potential for trace organic analysis in waste effluents. Christman (90) has applied this technique for analysis of lignin sulfonates in the waste effluents from Kraft

sulfite pulping operations. The author has discussed the analytical feasibility of *in situ* fluorescent spectrophotometry versus conventional analytical techniques. Spent-sulfite effluents from the Kraft process have
been monitored in concentrations as low as 0.2 mg/liter. Christman (90)
has also applied the same technique for analysis of organic color-forming
compounds in natural waters.

Traces of carbohydrate in the nanogram per liter range have been
determined in water samples using spectrofluorometry (386). The test
is based on the Seliwanoff reaction for the determination of ketones.
The sensitivity of the method and the effects of environmental factors
have been discussed (386).

The ensuing discussion of analytical methods for specific organic substances is subdivided into major groups of compounds (e.g., phenols,
detergents or surfactants, insecticides, etc.). Analytical procedures for
miscellaneous compounds found in wastewaters are listed in Appendix I.

### 4. Analysis for Phenols

Phenolic matter in industrial waste effluents is generally comprised
of monohydroxy derivatives of benzene and homologous and condensed
nuclei (20). Phenols in wastewaters, sometimes referred to as "tar acids,"
include cresols, xylenols, chlorophenols, naphthols, and other phenolic
derivatives. Phenols are found in a variety of industrial wastes (e.g.,
coal tar, gas liquor, plastic wastes, rubber-proofing wastes, cutting-oil
wastes, and the wastes from manufacture of disinfectants and
pharmaceuticals).

The discharge of phenolic wastes to a receiving stream is undesirable
because of the toxicity of this material to aquatic life, including fish.
Further, chlorination of a phenol-containing water for use as a water
supply results in the formation of chlorophenols and the resultant production of objectionable taste and highly intense odor.

One of the early methods for analysis of phenolic compounds in natural
and wastewaters was that of Fox and Gauge (154). The method is based
on coupling of the phenol, in caustic soda solution, with a freshly prepared
diazotized sulfanilic acid. Various shades of color are produced, depending on the particular phenols present. Generally, the color varies from
yellow to orange. For example, monohydroxy derivates of benzene give a
yellow color, mixed cresols give a yellowish-orange color, and xylenols
give a deep orange color. The reproducibility of this test can be improved
by carrying out the diazotization in ice-cooled solutions, using freshly
prepared diazonium compounds, and by properly controlling the pH of

the reaction. For waste samples containing interfering "tannins" (which contain phenolic groupings), chloroform extraction of the acidified solution will separate phenols from tannins.

Phenols in wastewaters from coke and gas works have been determined by a more sensitive photometric procedure based on substituting sulfanilic acid for the $p$-nitroaniline normally used in the Fox-Gauge test (53).

The more recent Gibbs method (168) for phenol has been applied extensively to natural waters and wastewaters (137,381). The method is based on the interaction of phenols with 2:6-dibromoquinone chloroimide and the formation of an indophenol dye, i.e.,

$$O:C_6H_2Br_2:NCl + C_6H_5OH \rightarrow O:C_6H_2Br_2:N\cdot C_6H_4OH + HCl \qquad (12)$$

Although the Gibbs method is more sensitive than the Fox-Gauge test, it also suffers from the formation of different shades of color for different phenolic compounds. For example, monohydroxy benzene and cresols give a blue color, $o$-cresol a purple color, and $p$-cresol gives no color at all. Amines interfere with the test, and sulfides prevent color development. Several modifications of the Gibbs method have been proposed to overcome some of these difficulties (7,200,336).

The standard method for the analysis of phenols in industrial wastewaters is based on the interaction of phenols with 4-amino antipyrine and oxidation under alkaline conditions to give an intense colored product. The reaction is dependent on the pH and type of buffer used (135a,135b). Variations of the pH value between 8.0 and 11.0 and the use of different buffer systems greatly influenced the sensitivity of the test to a particular phenol by shifting wavelength of maximum absorbance (144).

The reaction is also dependent on the type of phenol and the number of substitutions on the phenol ring (145a,145b). The position, type, and number of substitutions on the phenol ring were found to be greatly significant. Substitutions in the *para* position by alkyl, aryl, ester, nitro, benzoyl, nitroso, amino, and aldehyde appear to block or inhibit the reaction, the degree of which is dependent on pH. Faust and Mikulewicz (145a,145b) concluded that the determination of "total" phenol content in a water sample by the 4-amino antipyrine method is not possible since a number of phenols are either weakly reactive or do not react at all. The authors indicated, however, that best results are obtained when the test is done at pH 8.0 instead of 10.0 as recommended by *Standard Methods* (29).

Because the 4-amino antipyrine test cannot detect *para*-substituted phenols, other colorimetric procedures have been proposed for analysis for this substance (199,282). Water samples containing mixtures of phenol are first subjected to the 4-amino antipyrine test, then passed through an ion exchange column containing the hydroxide-form of Dowex 1-X8. The dye complexes pass through the column, while the para-substituted phenols remain sorbed. Elution with methyl alcohol, sodium chloride solution, and acetic acid removes the sorbed phenols, which are then diazotized with sulfanilic acid and determined spectrophotometrically at 495 m$\mu$. This method has been used for analysis for p-cresol, 2,3-xylenol, o,4-xylenol, and p-ethylphenol in various waters. Aromatic amines interfere with the reaction, and phenols containing para-substituted alkylaryl, nitro, and many other groups do not react. Oxidizing and reducing agents which react with the ferricyanide must be absent from the test solution. Despite these difficulties, the method is considered to be the best colorimetric test for most industrial wastewaters (121).

When phenols are present in fairly high concentration (e.g., more than 20 mg/liter), a titrimetric method based on bromination may be used (74). The waste sample is treated with excess standard bromine solution, which is then back-titrated iodometrically. The test is based on reaction

$$C_6H_5OH + 3Br_2 = C_6H_2Br_3 \cdot OH(\text{tribromophenol}) + 3HBr \qquad (13)$$

(13). The titrimetric method for phenols is susceptible to a variety of interferences from other organic species which may react with bromine.

Various modifications of the colorimetric procedure have been devised in attempts to reduce the effects of interferences. Double distillation techniques have been used to remove interferences from trade effluents (3). This has been followed by color development with 4-aminophenezone in an ammonium persulfate–alkaline borate solution. In other cases, the distillation of phenol was done after the water sample had been treated with $CuSO_4$. The distillate was then buffered to pH 8.3 and treated with alcoholic pyramidone and $K_3Fe$ $(CN)_6$ solution for color development (106). Phenols in waste effluents have been determined in the presence of resorcinol by taking advantage of the fact that dimethylamine-benzaldehyde (DBA), which is used to detect resorcinol, is insensitive to phenols. Because resorcinol interferes in an additive fashion, phenol in waste liquor samples can be determined by subtracting light absorption values obtained with DBA from those obtained with DBC (420).

Gas chromatography has been used successfully for analysis of a variety of substituted phenols which do not respond to the colorimetric test, at concentrations less than 1.0 mg/liter (38,144,351). Concentration by vacuum distillation and solvent extraction prior to chromatographic analysis have been reported (405). Paper chromatography has also been used for analysis of simple phenolic compounds in coal distillation waters (260), with n-heptane and cyclohexane being used to separate and clearly define spots of *meta-* and *para-*cresols.

Ultraviolet differential spectrometry has been used for analysis of phenolic substances in some instances (405). This technique is based on comparison of the B band (benzenoid bands) of alkaline forms of phenolic substances with those in neutral solutions. Differences in spectra can be detected and enable identification in the presence of interfering materials which mask direct spectra.

Near ir spectroscopy has been employed for analysis of phenolic derivatives in industrial wastewaters from oil refineries (21). The procedure is based on measuring ir spectra of the hydroxyl group stretching vibrations in monohydroxyphenols at 2.84 $\mu$ after bromination (420). The method is applicable to phenolic compounds which can be brominated in the *ortho* position with respect to the hydroxyl group. *Ortho*-substituted or sterically hindered phenolic compounds will not respond to this procedure. Polarographic determination of nitrochlorobenzenes in water using the standard addition technique has been reported (149). The method is based on separation on activated carbon and liquid extraction with acetone followed by extraction in $0.2M$ $C_5H_5N\cdot HCl$ in $C_4H_5N$ mixed 2:3 with 7% barium chloride. After removal of molecular oxygen by bubbling nitrogen and using alkaline pyrogallol solution, the polarographic wave is determined from 0.2 to 0.9 V versus an internal mercury electrode. The polarogram is repeated with a standard addition of nitrochlorobenzene.

## 5. Analysis for Pesticides

Pesticides are chemical substances employed for regulation of noxious fauna and flora. Although pesticides contribute significantly to the welfare of mankind, the indiscriminate use and discharge of these materials to the environment may constitute a very serious health hazard. There are approximately 300 organic pesticide chemicals, marketed in more than 10,000 different formulations (315). The national consumption in 1965 has been estimated to be about 750 million pounds of insecticides and herbicides (315). The points of concern in the use of pesticides are amounts of discharged toxicities, persistence, and rates of detoxification.

Chlorinated hydrocarbons, organophosphorus, carbamate insecticides, dithiocarbamate fungicides, dinitrocresols, and certain inorganic compounds are among the main types of pesticides. Pesticides are considered toxic in trace concentrations, and because of their persistence and cumulative effects, considerable emphasis has been placed on devising methods for separation and detection of trace quantities of these substances.

Liquid–liquid extraction (72,79) and adsorption on carbon (72,488) have been primary methods for the concentration and separation of pesticides from large volumes of natural waters and wastewaters. Methods of extractions vary; for example, many of the nonionic pesticides of the chlorinated hydrocarbon and organic phosphate classes can be extracted from either acid or basic solution with nonpolar solvents such as $n$-hexane. On the other hand, anionic compounds, such as phenoxyalkanoic acid herbicides, are extracted with a more polar solvent, from a solution which has been acidified to suppress the ionization of carboxylic groups. Quite often compounds are separated into nonionic and anionic fractions by adjusting the pH of the sample between extractions. Since large volumes of both water and solvent are often used for extraction, the solvent containing the residue must be reduced in volume or removed entirely in order to further concentrate the residue prior to separation and identification. Removal of solvent is often accomplished by evaporation under a current of air, by reflux distillation, or by vacuum evaporation.

Certain ionic compounds may require chemical modification to change their volatility prior to determination by gas chromatography. Burchfield and Johnson (79) have suggested treatment with diazomethane, reaction with a mixture of methanol and $BF_3$ and esterification with methanol and $H_2SO_4$.

There are four general methods for identification of pesticide residues: (a) spectrophotometric analysis, (b) elemental analysis, (c) measurement of biological activity, and (d) chromatographic analysis. One of the early colorimetric methods for the determination of concentrations of DDT as low as 0.003 ppm is the Schechter-Haller method (56,402). After preliminary isolation by extraction with a mixture of ether and $n$-hexane, the residue is nitrated to a polynitro derivative, a benzene solution of which gives an intense blue color with sodium methoxide dissolved in methyl alcohol. $\alpha$-Isomers of benzene hexachloride (gamexan) have been identified after extraction with carbon tetrachloride and dechlorination to benzene followed by nitration to $m$-dinitrobenzene. This last compound is then allowed to react with methylethylketone and alkali to give a reddish-violet color which may be measured photometrically (403).

Direct identification of pesticide residue after extraction and concentration has been done by infrared spectrometry (389) and by ultraviolet spectrometry (334).

Elemental analysis of a pesticide residue commonly is used to determine large amounts of certain compounds. The method relies on special properties which certain pesticides possess which are not shared by most naturally occurring metabolites, e.g., the organically bound chlorine in chlorinated hydrocarbons. If the organic fraction containing bound chlorine is combusted, the liberated chloride ions can be measured colorimetrically or titrimetrically (175). The total amount of pesticide will be expressed then in terms of gram-atoms of chlorine. Similarly, organophosphorus compounds can be determined by measuring the amount of phosphorus liberated. It must be realized, however, that this method not only lacks specificity, but also is subject to interferences from other chlorinated or phosphorus-containing compounds.

A more specific test for certain pesticides is to measure their antienzyme activity. Organophosphorus compounds are, for example, specific inhibitors for acetylcholinesterase. Methods are available for measuring the extent of enzymatic inhibition (382). These procedures have certain shortcomings, since different compounds exhibit different inhibitory action, and for the same phosphate content one may get different anticholinesterase activities. Fish bioassay (bluegill) has been used, however, to detect pesticide residues in the part per billion range (490). The test is based on measuring the inhibition of acetylcholinesterase in the brain of fish.

Chromatographic techniques probably provide the most sensitive procedure for the analysis of pesticides. Paper and thin layer chromatography (5,171,217,425) have been used separately or together with gas chromatography for pesticide analysis. An increasing number of articles are appearing in the literature describing new and more sensitive chromatographic techniques (72,79,130,255,296,334,483). Electron capture detectors have been used to detect as little as a picogram of chlorinated hydrocarbons and organophosphorus pesticides.

When the extreme sensitivity of the electron capture detector is not required, the microcoulometric titrator detector is preferred because of its high selectivity to halogens (except fluorides), sulfur, and phosphorus. With such high specificity the problems associated with the presence of impurities are minimized.

Identification of pesticides is usually based on chromatographic retention-time data coupled with the response of specific detectors. More confirmation is sometimes sought through correlating results from more than one chromatographic column or through comparing chromatographic data with data from thin layer or paper chromatography measurements. Comparison of infrared spectra is considered the ultimate identification procedure. Column eluents are collected, repetitively, and micropellets or microcavity cells are used for measuring the infrared spectra. Rapid scan infrared spectrometers connected directly to chromatographic columns have been used. Complete infrared spectra in the gas phase are obtained within 30 sec or less for the column effluent.

### 6. Analysis for Surfactants

As a result of the widespread use of synthetic detergents in domestic and various industrial applications, these materials are found commonly in sewage and industrial effluents. The objectionable tendencies of detergents to cause persistent foam in streams and at sewage plants and the possible toxicity of some of their constituents to fish and aquatic flora make it quite important to have suitable methods for their analysis.

Surfactants may be categorized as: (a) anionic, e.g., alkylbenzene sulfonate (ABS); (b) nonionic, e.g., polyglycol ethers of alkylated phenols (Lissapol N); and (c) cationic, e.g., quaternary organic bases such as roccal. Commercial detergents for domestic and industrial use are usually mixed with other substances, called "builders," which serve to improve the cleansing action of the detergents. Some of the common builders are sodium triphosphate or polyphosphate, sodium sulfate, sodium carbonate, sodium perborate, sodium silicate, and sodium carboxymethylcellulose. Anionic-type surface active agents have been employed most commonly for both domestic and industrial detergents. Within this class, alkylbenzene sulfonates are particularly difficult to degrade biologically during sewage treatment, and about 50% of the original amount commonly passes in the final effluent to receiving waters.

Under pressure from the public and State and Federal Governments, the detergent industry has introduced more readily biodegradable surfactants, e.g., sugar esters of linear aliphatic monocarboxylic acids, alkyl ether sulfates, and fatty acid ester sulfonates. The most common of the so-called "biodegradable" surfactants is the linear alkylate sulfonate (LAS).

The standard analysis for alkylbenzene sulfonates in industrial wastewaters (20,35) is based upon formation of a chloroform-soluble colored complex with methylene blue. The intensity of the blue color is measured photometrically at a wavelength of approximately 650 m$\mu$. Compounds which combine with the surfactant molecule to inactivate the sulfonate site may block the methylene blue reaction, thus causing negative interference.

In an attempt to minimize the effects of interferences, extraction of the methylene blue complex with chloroform in an alkaline solution (phosphate–sodium hydroxide buffer at pH 10) instead of an acid medium has been introduced (267). The chloroform extracts are eventually washed with an acid solution of methylene blue. By using this double extraction procedure, interferences due to chloride, nitrate, thiocyanate, and proteins are minimized.

Anionic surfactants have been determined by solvent extraction of the 1-methyl-heptylamine salt and measurement of color intensity spectrophotometrically at 650 m$\mu$. The effects of interferences in this procedure are reduced by double extractions, at pH 7.5 with a chloroform solution of 1-methyl-heptylamine and, after acid hydrolysis, extraction at pH 4.8 with a hexane solution of 1-methyl-heptylamine (143).

Separation by carbon adsorption has been used to avoid interferences (20,29). This technique, capable of detecting ABS in the parts per billion range, is based on (a) adsorption of the syndets on activated carbon, (b) desorption with alkaline benzene–methanol, (c) acid hydrolysis to destroy interfering organic sulfates, etc., (d) treatment with light petroleum to remove hydrocarbons, alcohols, and sterols, (e) extraction of surfactant in chloroform as a complex 1-methyl-heptylamine salt, and, finally, (f) infrared identification in carbon tetrachloride solution at 9.6 and 9.9$\mu$.

A number of review articles (192,261) discussing analyses for nonionic and anionic surfactants in waters and waste effluents have been published. Some of the more pertinent analytical procedures reported in the recent literature are listed in Appendix II.

## 7. Analysis for Combined Nitrogen
### (Ammonia, Nitrites, and Nitrates)

$NH_3$, $NO_2^-$ and $NO_3^-$ in wastewaters may result from the degradation of organic nitrogenous compounds or may be entirely of inorganic origin.

The most widely used method for analysis for ammonia is the Nesslerization reaction. The test is based on the development of a yellow–brown (colloidal) color on addition of Nessler's reagent to an ammonia solution. The standard method (29) and the ASTM reference test (20) recommend the separation of the ammonia from the sample by distillation prior to the Nesslerization reaction. Direct Nesslerization is most often preferred, however, for rapid routine determinations.

For certain industrial wastewaters, it is often desirable to distinguish between "free" ammonia and "fixed" ammonia. The former is estimated by a straightforward distillation; the residual liquor is then treated with excess alkali (e.g., sodium carbonate, magnesium oxide or caustic soda) and distilled to determine fixed ammonia. Certain substances interfere with both the direct Nesslerization and distillation mtehods, e.g., glycine, urea, glutamic acid, acetamides, and hydrazines.

The standard method (20,29) for nitrites in water is based on forming a diazonium compound by the diazotization of sulfanilic acid by nitrite under strongly acidic conditions and coupling with α-naphthylamine hydrochloride to produce a reddish-purple color. Spectrometric measurement of the color of the azo dye is performed at 520 mμ, or visual comparison with standards may be used. This method is sometimes known as the Griess-Ilosvay method (379). A frequently used alternative procedure for nitrite is based on formation of a yellowish-brown dye by the reaction of nitrite in acid solution with *meta*-phenylene diamine (238).

The Griess-Ilosvay method is most suitable for low nitrite concentrations, e.g., below 2.0 mg/liter. Another colorimetric procedure which is more suitable at high concentration involves reaction of the nitrites with *meta*-phenylene diamine in an acid solution to form a yellowish-brown dye. Chlorides in concentration below 500 mg/liter have no effect on either of the two procedures. High chloride concentrations (10,000 mg/liter) interfere more with the Griess-Ilosvay test. More distinct, pH-independent color development is achieved (418) by replacing the sulfanilic acid with sulfanilamide, and α-naphthylamine with 1-naphthylethylene diamine dihydrochloride within the Griess-Ilosvay test. The sulfanilamide undergoes diazotization in hydrochloric acid and the diazonium salt is then coupled with the diamine to give a stable red azo dye.

An excellent review of approximately 52 spectrophotometric methods for nitrite has been published by Sawicki et al. (399). The authors critically evaluate the sensitivity, color stability, conformity to Beer's law, simplicity and precision of a variety of methods.

Colorimetry, uv spectrometer, and polarography have frequently been used for nitrate determinations in natural waters and waste effluents. The phenoldisulfonic acid method and the brucine method are two colorimetric procedures more frequently used. In the former test, color development is based on the reaction between phenol disulfonic acid and nitrates in sulfuric acid solution to give a nitro derivative which causes a yellow coloration when the solution is made alkaline; the intensity of the color is measured at 470 mμ. Nitrite ion interferes with the test in proportion to its concentration in the sample. Various inorganic ions above certain concentrations cause interference (20). Small amounts of chlorides do not interfere, but nitrites should be removed with sodium azide (20).

An alternate test for nitrates involves reaction of a brucine solution in glacial acetic acid with nitrates and acidification with dilute $H_2SO_4$. The color intensity changes with time and it is necessary to develop the color of standards and samples simultaneously and compare maximum color intensity. Chlorides above 1000 mg/liter interfere with color development. Nitrites, if present, should be separately estimated and an appropriate correction applied. A salt-masking technique which renders the test applicable to sea water and brackish water has been proposed by Jenkins and Medsker (219).

Nitrate analysis by reduction to ammonia, which is then detected by Nesslerization, has been reported by several authors. The procedure is based on expelling all ammonia from the water sample, followed by reduction of nitrogen ($NO_2^-$ and $NO_3^-$) by means of (a) aluminum foil in alkaline NaOH solution, (b) zinc–copper couple in acetic acid solution, (c) Devarda's alloy hydrazine (93), and (d) alkaline ferrous sulfates. The ammonia produced may be separated by steam distillation and estimated in the distillate by Nesslerization. Various procedures have been proposed to minimize interferences due to nitrites and chlorides.

Nitrate analysis by reduction to nitrites which are then detected by the Griess-Ilsovay method has been applied to both natural waters and wastewaters (127). Controlled reduction of nitrates to nitrites is accomplished with zinc powder in acid solution.

Ultraviolet analysis for nitrates offers the advantage of freedom from chloride interferences and a variety of other inorganic ions. However, dissolved organic compounds, nitrites, hexavalent chromium, and surfactants interfere with this procedure. The test is based on measuring uv adsorption spectra of the filtered, acidified sample at 220 m$\mu$. Measurements follow Beer's law up to 11 mg N/liter. Interference of dissolved organics is estimated by doing a second measurement at 275 m$\mu$, a wavelength at which nitrates do not absorb.

Simultaneous determination of nitrates, nitrites, and sulfates in water samples by infrared techniques has been reported (93). The test is based on concentrating the sample by ion exchange and removal of phosphates, carbonates, and organic matter. This is followed by separation by freeze-drying the aqueous solution in the presence of KBr; the infrared spectrum is determined in the resulting KBr disk.

The polarographic analysis of nitrates in wastewaters is based on the original work of Kolthoff, Harris, and Matsuyama (241a). Nitrate ion is catalytically reduced at the dropping mercury electrode in the presence of uranyl ion in an acid solution at $-1.2$ V versus SCE. The diffusion current is linearly proportional to the nitrate ion concentration. Nitrites, phosphates, ferric iron, and fluorides interfere with the test. Procedures to minimize interferences have been prescribed. The polarographic test offers the advantage of being adaptable to continuous monitoring (20,29,156).

The analytical procedures described above for ammonia, nitrites, and nitrates are those most commonly applied to wastewaters. Other pertinent procedures reported during the last few years are listed in Appendix III.

### 8. Analysis for Combined Phosphorus
### (Ortho-, Pyro-, and Polyphosphates)

Phosphorus may be present in industrial waste effluents either as inorganic phosphates (ortho-, meta-, or polyphosphates) or in organic combination. The most common analytical method for inorganic phosphorus is based on the colorimetric determination of the phosphomolybdenum blue complex (20,29). The test is sensitive to orthophosphates and not condensed phosphates. Polyphosphates and metaphosphates are then estimated as the difference between total phosphates (hydrolyzed samples) and orthophosphates (nonhydrolyzed samples).

Orthophosphates react with ammonia molybdate in acid medium to form the phosphomolybdic acid complex, which when reduced yields the molybdenum blue color which may be determined colorimetrically. The sensitivity of the test is largely dependent on the method of extraction and reduction of the phosphomolybdic acid aminonaphthol-sulfonic acid (29), stannous chloride (Deniges method) (6), metal sulfites (Tschopp reagent) (463), and ascorbic acid (150) have been used in the reduction step. The stannous chloride method is considered most sensitive and best suited for lower ranges of phosphate concentration.

A number of substances have been reported to interfere with the phosphate determination (29). Arsenic, germanium, sulfides, and soluble iron above 0.1 mg cause direct interferences. Tannins, lignins, and hexavalent chromium will cause errors only for analysis of phosphate concentrations below milligrams per liter.

Various modifications of the above procedures, as well as other new techniques, have been recently reported; some of the more pertinent ones are given in Appendix III.


## C. BIODEGRADABILITY OF ORGANIC COMPOUNDS

With increasing emphasis on water pollution control, and in view of recent legislation restricting the disposal of waste effluents, biodegradability is becoming the most significant test in the analysis for industrial wastewaters. Biodegradability tests are basically designed to estimate the extent to which organic compounds may be oxidized biochemically, and an industry must be concerned with the question of whether its products or waste materials can be degraded or assimilated efficiently by existing biological waste treatment processes.

Biodegradability may be considered as a measure of the susceptability of organic material to microbial metabolism. This property is not well defined and there is no single standard test for its measurement. It constitutes, however, a dominant mechanism for the removal of organic pollutants from water, both in self-purification processes in natural waters and in the accelerated biological processes of waste treatment. Biodegradation can occur aerobically or anaerobically, depending on the availability of atmospheric oxygen. The process is affected by a variety of environmental conditions. As discussed previously in this chapter, organic compounds are not equally susceptible to biodegradation; some are readily metabolized and others are more resistant (refractory compounds). Principal factors involved in any biodegradation process are as follows.

*a.* Type and number of microorganisms: a mixed culture of organisms, as in sanitary sewage, possesses a remarkable capacity to adapt to strange or different organic materials, while single cultures may not be effective.

*b.* Structure and concentration of organic materials: certain organic compounds, e.g., certain pesticides and surfactants, are relatively resistant to biodegradation in comparison to simple carbohydrates. The concentration of the organic material is also significant. High concentrations of "sugars" in certain wastewaters may inhibit biodegradation, yet upon dilution in a receiving water the sugars will be easily degraded, resulting in a water pollution problem due to oxygen depletion.

*c.* Environmental factors: factors such as temperature, mixing, and viscosity, of the wastewater are quite significant in dictating the extent and rate of biodegradation and its effect on the ecology of the receiving environment (air, water, or soil).

Much of the work which has been carried out on biodegradation has been concerned with surfactants and, in particular, with the "hard" branched-type ABS and the "soft" straight-chain LAS (18,26,27,301, 426,445,467). Surfactant (ABS or LAS) biodegradation begins at one end of the alkyl chain with oxidation of the terminal methyl group to a carboxyl group. Beta oxidation follows, shortening the chain by two carbons at a time. Then the benzene ring splits, yielding unsaturated intermediates (301,445). The rate of biodegradation is dependent on the degree of branching of the paraffin part of the surfactant molecule (301). Because of its linear configuration, linear alkylate sulfonate (LAS) degrades at a faster rate and requires less residence time in the biological treatment plant than does ABS.

A number of techniques have been used to test for biodegradability, such as the Warburg respiration technique (468), activated sludge tests (26,444), shake-flask experiments (444), and river die-away measurements (18,180). A comprehensive critique of various biodegradability tests has been presented recently by Bunch and Chambers (78). In some cases, investigators have tried to devise procedures to duplicate as closely as possible the conditions prevailing in the receiving stream or in the waste treatment plant (18,301). In other cases, simple static procedures have been employed (78). Sewage microorganisms are commonly used for biodegradability measurements because of availability and to avoid the need for maintaining standard cultures (78). Biological oxidation has been followed by measurement of one or more of the following parameters: (a) the rate of disappearance of the organic compound under test, (b) the rate of appearance of biodegradation by-products, (c) growth rate of microorganisms, and (b) dissolved oxygen consumption.

Analytical techniques used for estimating biodegradability have been reported by Allred et al. (18) and Bunch and Chambers (78). A number of laboratory-scale test environments have been proposed for measuring surfactant biodegradability (26).

Bunch and Chambers (78) have described a static biodegradation test and its application to a number of organic compounds. Gates et al. and Mancy (161,283) have investigated the mechanism and rates of biochemical assimilation of organic compounds in small laboratory batch reactors. Variable levels of turbulence were rated in terms of values of air–water oxygen transfer coefficients. The results point out the effect of the type and concentration of organic compounds and the type and number of microorganisms and assimilation rates.

## VI. ANALYSIS FOR METALS

This section is concerned with the analysis of metal ions commonly found in industrial wastewaters. The metals of interest include alkali metals, alkaline earths, transition metals, and heavy metals. Depending on the physicochemical characteristics of the solution phase in which they are dissolved, metal ions may occur in one or more aquometallic or organometallic complexes.

Metal analyses have undergone significant changes in the last three decades. Prior to about 1940, most analytical techniques for metals were either gravimetric or volumetric. Since the post-War era of the late 1940's there has been a considerable increase in the use of the spectrophotometric techniques, largely as a result of the development of various organic reagents such as dithizone, o-phenanthroline, sodium diethyldithiocarbamate, and diphenylcarbazide, which form color-producing compounds with metal ions in solution. Many of these reagents are highly specific for particular metals and find wide application in water analysis. Compleximetric titrations with ethylenediamine tetraacetate (EDTA) and a variety of specific metal ion indicators also are used extensively for analysis of metal ions in waters and wastewaters. Perhaps the most familiar example is the analyses for calcium, magnesium, and water hardness by titration with EDTA (29).

## A. SEPARATION AND CONCENTRATION TECHNIQUES

Various separation and concentration techniques are available for removing interferences, for extraction of colored organometallic complex compounds, or for concentration prior to titrimetric, spectrometric, electrometric, or radiometric analysis. In certain cases, the separation itself is sufficiently specific that it may be followed by a nonselective analytical procedure. In other cases, a mixture of two or more metals ions may be separated from solution and then subjected to analysis by more selective analytical procedures.

Separation by ion exchange has been used extensively for metal cations in natural and wastewaters. Perhaps the first analytical application of organic ion exchange was Kullgren's work on the separation of copper ions from wastewater for subsequent determination (244). The total free metal ion content of a wastewater can be determined by ion exchange (98,214,215). The technique involves passing a sample of water or waste through a hydrogen-form cation exchange resin and titrating the equivalent quality of $H^+$ released with a standard base. Another aliquot may be titrated with EDTA for the hardness metals. Analyses for $NH_4^+$, $K^+$, $Rb^+$, and $Cs^+$ ions in water have been carried out by precipitation with tetraphenylboron. The precipitate is dissolved in acetone and the solution passed through a cation exchange resin in the hydrogen form; the resulting free tetraphenylboron is then titrated (215).

Ion exchange also can be used effectively for removal of interfering ions or cations from a water sample. For example, the separation of complex cyanide ions which interfere with the titrimetric determination of alkali metals can be achieved by an ammonium-form cation exchange resin. This method also separates other interfering anions, such as vanadates, chromates, molybdates and tungstates (214). McCoy has presented an interesting discussion of the use of ion exchange for total separation of various anions and cations from industrial wastewaters (297).

Ion exchange chromatography offers an effective method for concentration and separation of ions from wastewaters. The ions are first concentrated on a suitable ion exchange column and then selectively eluted to be determined polarographically, radiometrically, spectrophometrically, or spectrographically (126,178,302). Iron, commonly found in industrial waste effluents, may be collected on cation exchange columns as ferric iron, reduced to the ferrous state by a dilute ascorbic acid solution, and then eluted with a strong acid (214). The selective separation of metals by ion exchange chromatography often can be markedly improved by using complexing agents (158). For example, chromatographic separation of alkaline earth (magnesium, calcium, strontium, and barium) ions from a cation exchange column can be accomplished by elution with hydrogen chloride, ammonium acetate, ammonium formate, ammonium lactate, EDTA, diaminocyclohexanetetra-acetate, or ammonium -oxy-isobutyrate (214). The order of separation of ions is in all cases the same regardless of which of these complexing agents is used.

Attempts have been made recently to use ion exchange membranes for the separation and concentration of metal ions prior to analysis (131,283). The technique involved mounting a cation exchange membrane, of 2–5 mils thickness, on the surface of an indicator electrode (usually carbon, platinum or gold) in a voltammetric system. The membrane serves as an ion exchange preconcentration matrix, as well as a rigorously defined diffusion barrier for surface active or electroactive interferences present in the test solution. The exchange rate of metal ions between the test solution and membrane is accelerated by applying an appropriate emf across the membrane.

This technique was illustrated by Eisner and co-workers (131), who used a membrane-covered carbon electrode in $10^{-4}$ and $10^{-5}$ $M$ silver ion and 0.1 $M$ potassium nitrate solutions. Separation and concentration of the silver was done by potentiostating the membrane electrode at $-0.4$ $V$ versus SCE for 4 min; then the deposited metal was stripped by applying a 10 mV/sec anodic sweep. Well defined voltammetric peaks were obtained with a definite gain in sensitivity. When egg albumin at a concentration of 0.01% was added to the silver ion solution, the sensitivity obtained was ten times as great as that with the ion exchange membrane mounted on the electrode surface. In this case the membrane, being practically impermeable to protein molecules, served as a protective diffusion barrier for the indicator electrode.

Perhaps the most interesting aspect of ion exchange membranes in this application is their suitability for *in situ* separations. Hence, membrane-covered electrode systems can be used for analysis of metal ions in their natural environment. Additionally, the ion exchange membrane itself can be used as a preconcentration matrix for subsequent determinations by activation analysis, emission, or absorption spectrophotometry.

The main problem associated with using cation exchange membranes for transition and heavy metals in natural waters and wastewaters is their lack of specificity. This is particularly significant since the alkalies and alkaline earths are usually present in great excess. Recent exploratory studies (283) demonstrated the feasibility of using chelate ion exchange membranes made from Chelex 100. Such ion exchangers showed high selectivity for multivalent cations.

Another limitation to the use of cation-exchange membranes for natural waters and wastewaters is the fact that, except for the alkalies and alkaline earths, most metal ions will be found in the form of organometallic complexes. The separation of such complexed metal ions by ion exchange is sometimes not possible unless the complex is first disrupted.

Solvent extraction is commonly used for separation of metal ions from industrial wastewaters. Valuable schemes for the separation of a large number of metal ions by successive extractions employing different complexing agents and organic solvents at controlled pH have been reported (86,252,280,305,409,422,494). These discussions cover the colorimetry, sensitivity, and selectivity of principal reagents, e.g., dithizone (diphenylthiocarbazone), oxine, cuperferrons, diethyldithiocarbamate, tetraphenyl arsenium salts, quaterary ammonium compounds, and various chelating agents.

Selection of an appropriate solvent–extraction system depends on its specificity and its suitability for subsequent analytical procedures. By selection of appropriate ion-association and chelate solvent systems, almost any element can be separated from wastewaters. Examples of solvent extraction systems which have been or could be used for the analysis of industrial wastewaters are given in Table VII.

Partial freezing also has been used as a technique for cation concentration (239,256,281,415). The procedure involves the slow freezing of a water sample, the time required to achieve a certain concentration usually

TABLE VII

Solvent Extraction Systems Applicable to Wastewater Analysis

| Elements | Sample | Complexing agents | Solvent | Method of analysis | Ref. |
|---|---|---|---|---|---|
| Ce | Biological matter | Bis(2-ethyl hexyl)-hydrogen phosphate | n-Heptane | Radiochemical | 181 |
| Cr$^{VI}$ | Sea water Brines | Dithizone | Methyl isobutyl ketone (MIBK) | Atomic absorption | 113 |
| Cu | Sea water | Dithizone | CCl$_4$ | Polarographic | 459 |
| Cu | Sea water | Sodium diethyldithiocarba-mate (DEDC) | CCl$_4$ | Spectrophoto-metric | 441 |
| Cu | Sea water | 2,2'-Diquinolyl neocuproine | n-Hexanol | Spectrophoto-graphic | 268 |
| Fe | Sea water | Diphenylphenanthroline | Isobutyl alcohol | Spectrophoto-metric | 417 |
| Mn | Sea water | 1-Nitroso-2-naphthol | CHCl$_3$ | Spectrophoto-metric | 268 |
| Mo | Brines | Dithiol | MIBK | Atomic absorption | 113 |
| Pb | Environmental samples | NaI, dithizone | Isopropyl methyl ketone | Radiochemical | 448 |
| Zn | Urine | Dithizone | CCl$_4$ | Spectrophoto-metric | 226 |
| Co, Ni | Natural waters | Ammonium pyrrolidine | MIBK | Atomic absorption | 81 |
| Co, Zn | Sea water | Dithizone | CCl$_4$ | Spectrophoto-metric | 152 |
| Fe, Mn | Sea water | DEDC | MIBK | Atomic absorption | 223 |
| Cu, Mo, Zn | Sea water | 8-Quinolinol | CHCl$_3$ | Spectrochemical | 76 |
| Cu, Mo, Mn, V | Brines | APDC | MIBK | Atomic absorption | 289 |
| Cu, Fe, Mn, Ni | Brines | Cupferron | MIBK | Atomic absorption | 113 |
| Fe, Mn, Mo, Zn | Milk | APDC | CHCl$_3$ | Emission spectro-graphic | 480 |
| Ag, Cu, Fe, Mo, Ni, V | Sea water | DEDC | CHCl$_3$ | Emission spectro-graphic | 515 |

506

being judged by experience. The cation recovery by this technique is given by the following relationship:

$$\% \text{ recovery} = \frac{C_{L2}/C_{L1}}{V_1/V_2} \times 100 \tag{14}$$

where $C_{L2}$ and $C_{L1}$ are the initial and final cation concentrations in the liquid, respectively, $V_1$ is the initial sample volume, and $V_2$ is the final volume of liquid residue.

Partial freezing has been used for the concentration of Fe, Cu, Zn, Mn, Pb, Ni, Ca, Mg, and K in water samples in concentrations ranging from 0.1 to 10.0 mg/liter (281). Increasing the mixing rate up to some limiting value increases cationic recovery. The effect of initial pH on recovery efficiency depends on the nature of the cation. Alkali metals (K, Ca, and Mg) concentrate best at low pH, while heavy metal cations (Pb, Ni, and Cu) concentrate best under alkaline conditions (281).

## B. INSTRUMENTAL METHODS

### 1. Absorption Spectrophotometry

Determination of metal ions in wastewaters by absorption spectrophotometry is based primarily on reaction of the metal ions with various organic reagents to form colored compounds which may be determined spectrophotometrically either directly or after appropriate separation. A compleximetric reaction between the metal ion and the organic molecule—acting often as a multidentate ligand—is usually involved. Some of the more common organic reagents used for separation by extraction are chelate compounds, e.g., dithizone (diphenylthiocarbazone), oxines, cupferron, and diethyldithiocarbamate (82,86,153,323,381,382,396). Examples of applications of this technique for analysis of metals in wastewaters are given in Table VIII.

The sensitivity of the test may be increased by selection of appropriate reagents, wavelength, and type and length of cuvette. The basic law of absorption spectrophotometry, the Beer-Lambert law, relates the absorbance, $A$, directly to the concentration of the absorbing species, $C$, the length of the light path through the absorbing solution, $l$, and the molar absorptivity of the absorbing species, $\epsilon$, i.e.,

$$A = \log (I_0/I) = \epsilon l C \tag{15}$$

where $I_0$ and $I$ are the intensities of the incident and emitted light, respectively. It would appear from eg. (15) that simply increasing $l$ for a

## TABLE VIII

### Examples of Molecular Absorption Spectrophotometry for Metals in Wastewater

| Metal | Complexing agent | Solvent extraction | Color of complex | pH range | Suitable wave-length, mu | Useful range mg/liter | Refs. |
|---|---|---|---|---|---|---|---|
| Cobalt | Diethyldithiocarbamate | Ethylacetate | Blue | pH:3.0 | 367 | — | 20, 29 |
| Cadmium | Dithizone | Carbon tetrachloride | Red | pH:10-12 | 518 | 0.1-5 | 20, 29 |
| Chromium | 1.5-Diphenyl carbohydrazide | Butanol | Violet | pH:2-3 | 540 | 0.05-0.5 | 20, 29 |
| Copper | Dithizone | Carbon tetrachloride | Violet | pH:0.5 | 510 | 0.04-14 | 37 |
| Copper | Diethyldithiocarbamate | Carbon tetrachloride | Yellow-brown | pH:9.0 | 436 | 0.1-0.8 | 20, 29 |
| Copper | Cuprione (z-z'di-quinolyl) | Isoamyl alcohol | Purple | pH:5-6 | 540 | — | 382 |
| Iron | 1,10-Phenanthroline | — | Orange-red | pH:2-9 | 490 | 0.1-1.0 | 20, 29 |
| Iron | Thioglycollic acid | — | Purple | pH:8-12 | 540 | 0.04-1.2 | 20, 29 |
| Iron | Tripyridyl | — | Red-purple | pH:9-10 | 560 | 0.01-2.0 | 323 |
| Lead | Dithizone | Chloroform | Red | pH:7-10 | 520 | — | 20, 29 |
| Mercury | Dithizone | Carbon tetrachloride | Yellow-orange | pH:0-1 | 500 | — | 20, 29 |
| Nickle | Dimethylglyoxime | — | Reddish-brown | pH:9.5-11 | 465 | — | 20, 29 |
| Zinc | Dithizone | Chloroform | Purple-red | pH:4-5.5 | 530 | 0.1-1.0 | 20, 29 |
| Zinc | Zincon | — | Blue | pH:9.0 | 620 | 0.1-2.4 | 396 |

given concentration will increase the sensitivity of an absorption measurement proportionally. This is reasonably true for small changes in $l$, but the background absorption by the other reagents in solution becomes limiting for very large light path lengths. Increasing the sensitivity of such determinations depends ultimately on the formation of color compounds of high molar absorptivity.

Typical sensitivities of metal analyses by absorption spectrophotometry are about $10^{-6}$ $M$. It is possible in certain cases to exceed normal limits of sensitivity by using "amplification reactions," which often result in a severalfold increase in sensitivity, e.g., the determination of phosphates in the nanogram range (469). Differential spectrophotometry (86) also allows for much more precise determinations than are possible by conventional techniques. Molecular absorption spectrophotometric procedures generally can be automated readily.

Molecular fluorescence spectrophotometry offers certain advantages over molecular absorption spectrophotometry with respect to selectivity and sensitivity. The technique is based on the spectral measurements of fluorescence or phosphorescence radiation emitted from luminescent compounds upon excitation by incident radiation. The reemitted radiation is of lower frequency than the absorbed light. Fluorescence spectra are characteristic of the compound in the sense that the emission spectrum is always the same irrespective of the wavelength of the incident light which promotes the fluorescence.

The fluorescence equation may be expressed as follows:

$$F = [2.303\phi I_0 \epsilon l p]C \tag{16}$$

where $F$ is the amount of fluorescence generated, $\phi$ is a constant related to the efficiency of fluorescence, $I_0$ is the intensity of incident radiation, $\epsilon$ is the molar absorptivity at a given wavelength, $l$ is the pathlength in centimeters, $p$ is a fractional constant, and $C$ is the concentration. Hence, $F$ measured in terms of the signal response of a photomultiplier tube sensitive to fluorescence radiation is proportional to the analytical concentration $C$, while the parameters $I_0$, $l$, and $p$ are instrumental factors, and the parameters $\phi$ and $\epsilon$ are functions of the efficiency of the fluorescent reagent system.

It is evident from eq. (15) that for absorption spectrophotometry any increase in $I_0$ will be accompanied by a matching increase in $I$, with no net gain in the absorbance, $A$. However, for fluorescence spectrophotometry, any increase in $I_0$ will be matched by a corresponding increase in the analytical signal $F$, as indicated in eq. (16). Also, any increase in the amplifier gain in absorption spectrophotometry will amplify $I_0$ and $I$ correspondingly, whereas in fluorescence spectrophotometry, this will result in an increase in $F$. For these reasons the sensitivity of molecular fluorescence spectrophotometry is inherently greater than that of molecular absorption spectrophotometry.

Despite the fact that molecular fluorescence spectrophotometry has not been used to a very large extent for inorganic analysis, it offers an extremely useful technique which is applicable to solutions 100–10,000 times more dilute than those which can be analyzed by absorption spectrophotometry. There are a number of fluorometric reagents suitable for analysis of such metals as aluminum, rare earths, zinc, and calcium, which largely form colorless complexes. In those cases where the metals form colored complexes, such as iron, copper, nickel, and chromium, measurements are based on the extinction of the fluorescence of the reagents with which they react. A typical example of an effective fluorogenic agent is 8-hydroxyquinoline, which forms fluorescent complexes with aluminum, beryllium, etc., and nonfluorescent complexes with iron, copper, etc.

One of the most desirable characteristics of molecular fluorescence spectrophotometry for analysis of wastewaters is its selectivity. Only certain ions are capable of producing fluorescence. For example, few metal ions produce fluorescence with a non-selective reagent like 8-hydroxyquinoline, while over 30 produce absorption spectra.

Aluminum has been determined in natural and wastewaters by simple fluorometric techniques (120) using pentachrome blue–black to form a fluorescent complex. Trace quantities of aluminum have been determined by fluorometric techniques using 2-hydroxyl-3-naphthoic acid (234) and 2-pyridylazo-2-naphthol (443) as fluorogenic agents. Comprehensive discussions of fluorimetric analysis have been provided by White and Weissler (497,498) and West (494).

Perhaps the most dramatic achievement in inorganic analysis during the last decade has been the development of atomic absorption spectrophotometry. The similarity of atomic absorption spectrophotometry to molecular absorption spectrophotometry is based on the fact that atoms are capable of absorbing light in exactly the same way as molecules by interacting with a photon of the energy requisite to promote an electronic transition from ground state to one of the excited states of the atom. Hence the laws which govern the relationship between the amount of light absorbed and the concentration of the absorbing species, as well as the experimental apparatus and techniques, are basically the same for both atomic and molecular absorption spectrophotometry. As an analytical procedure, atomic absorption spectrophotometry has the unique advantage of virtual specificity. Exceptions are those few cases in which unfavorable matrix components are present in the sample solution. This is largely a result of the presence of certain compounds which combine with the metal under analysis to form relatively nonvolatile compounds, which do not break down in the flame. Calcium in the presence of phosphate exhibits this effect (55). This may be remedied by sequestering the calcium ion with EDTA. Matrix effects may be minimized by separation or by adding approximately the same amount of matrix component to the standard solutions.

In contrast to flame photometry, there is very little interelement interference in atomic absorption spectrophotometry. Also, while sensitivity in flame photometry is critically dependent on flame temperature, this is not the case for atomic absorption spectrophotometry.

Two excellent monographs on atomic absorption have been prepared recently by Robinson (383) and Elwell and Gidley (135). Review articles by Kahn (228), Fishman and Midget (148), and Boettner and Grunder (63) offer comprehensive surveys of applications in natural waters and wastewaters.

Over 60 elements can be determined readily by atomic absorption in the parts per million range without sample pretreatment and with an accuracy of ±1–2%. This sensitivity can be vastly increased to the parts per billion range by scale expansion or by extracting the metal in a nonaqueous solvent and spraying it into the flame. Microgram per liter quantities of cobalt, copper, iron, lead, nickel, and zinc have been determined in saline waters by extraction of metal complexes with ammonium pyrrolidine dithiocarbomate into methyl isobutyl ketone (148). The use of organic solvents may alter the flame temperature, which, in contrast to flame photometry, will generally have no significant effect. An increase of about 60% in the atomizer efficiency can be achieved with the use of certain organic solvents.

Although atomic absorption spectrophotometry is a relatively new technique, it is being applied widely for analysis of metal ions in natural waters and wastewaters. In addition to its selectivity and sensitivity, atomic absorpton spectrophotometry is a rapid and easy technique, suitable for routine analysis and easily automated for monitoring effluent streams (24,55). The detection limits to some common metals were reported by Kahn (228) and are given in Table IX.

From the above discussion, it would appear that molecular fluorescence spectrophotometry offers the advantages of greater sensitivity and selectivity over molecular absorption spectrophotometry. Relative to atomic

TABLE IX
Atomic Absorption Detection Limits[a]

| Metal | Detection limit | Analytical wavelength | Suggested resolution, Å |
|---|---|---|---|
| Silver | 0.005 | 3281 | 7 |
| Aluminum[b] | 0.1 | 3093 | 2 |
| Arsenic[c] | 0.1 | 1937 | 7 |
| Boron[b] | 6.0 | 2497 | 7 |
| Barium[b] | 0.05 | 5536 | 4 |
| Beryllium[b] | 0.002 | 2349 | 20 |
| Bismuth | 0.05 | 2231 | 2 |
| Calcium | 0.002 | 4227 | 13 |
| Cadmium[c] | 0.001 | 2288 | 7 |
| Cobalt | 0.005 | 2407 | 2 |
| Chromium | 0.005 | 3579 | 2 |
| Copper | 0.005 | 3247 | 7 |
| Iron | 0.005 | 2483 | 2 |
| Mercury | 0.5 | 2537 | 20 |
| Potassium | 0.005 | 7665 | 13 |
| Lanthanum[b] | 2.0 | 3928 | 4 |
| Lithium | 0.005 | 6708 | 40 |
| Magnesium | 0.0003 | 2852 | 20 |
| Manganese | 0.002 | 2795 | 7 |
| Molybdenum | 0.03 | 3133 | 2 |
| Sodium | 0.002 | 5890 | 4 |
| Nickel | 0.005 | 2320 | 2 |
| Lead | 0.03 | 2833 | 7 |
| Antimony | 0.1 | 2175 | 2 |
| Selenium[c] | 0.1 | 1961 | 20 |
| Silicon[b] | 0.1 | 2516 | 2 |
| Tin[c] | 0.02 | 2246 | 7 |
| Tellurium | 0.1 | 2143 | 7 |
| Titanium[b] | 0.1 | 3643 | 2 |
| Thallium | 0.025 | 2768 | 20 |
| Vanadium[b] | 0.02 | 3184 | 7 |
| Tungsten[b] | 3.0 | 4008 | 2 |
| Zinc | 0.002 | 2138 | 20 |
| Zirconium[b] | 5.0 | 3601 | 2 |

[a] The detection limit is given by the metal concentration in parts per million, which gives a signal twice the size of the peak to peak variability of the background.
[b] Nitrous oxide flame required.
[c] Indicates use of argon–hydrogen flame.

absorption spectrophotometry, however, no increase in selectivity can be gained by using atomic fluorescence, because the former is virtually specific for each element. It is possible, however, to increase the sensitivity of measurements with atomic fluorescence spectrophotometry by increasing the intensity of irradiation or by increasing the amplification until the system becomes noise limited (133). In this sense, atomic fluorescence spectrophotometry offers greater flexibility and sensitivity than atomic absorption spectrophotometry (347,505). The technique is inherently simple, and practically any flame spectrophotometer may be adapted for this purpose without interference with its normal mode of operation. A continuous source with simple monochromator may be used to readily generate atomic fluorescence. Where high sensitivities in the subnanogram range are required, it is necessary to use individual spectral discharge lamps (133).

## 2. Emission Spectrometry

Flame photometry has been used extensively for determination of alkaline metals and certain alkaline earths because of their low excitation energies. Reproducible results are obtained with careful control of flame temperature and sample composition and the use of internal standards.

Some of the main limitations of flame photometry are interelement interference (e.g., K and Mn) and the dependency of the energy of emission on the temperature of the flame. In some cases the temperature of the flame may be a limiting factor in determining whether or not the metal can be detected. With high temperature flames, such as those of $C_2H_2$—$O_2$ or $C_2H_2$—$N_2O$, up to 42 elements can be determined. The flame temperature, shape, background, and rate of sample consumption are critical factors controlling the sensitivity.

The majority of literature on the use of flame photometry for water analysis is concerned with sea water (88), river waters (220), and mineral waters (98). A number of investigators also have applied flame photometry to analysis of oil field effluents (98) and brine wastewaters (63). For certain applications separation by ion exchange has been used prior to determinations by flame photometry to remove anion and cation interferences or for concentration of certain elements (24,98).

With the advent of highly selective atomic absorption spectrophotometry, the applications of flame photometry to natural waters and wastewaters have been limited to a few alkalies and alkaline earth metals.

Emission spectroscopy is not frequently used for analysis of metals in wastewaters because of the highly specialized training required for proper operation and for interpretation of spectra. Further, few laboratories have the required instrumentation. It is often necessary to subject the water sample to separation and/or concentration procedures prior to the spectrographic analysis. Spectrographic procedures have been developed for routine analysis of 19 elements in river and ground waters (242) using rotating graphite electrodes and high voltage spark excitation. Wastewaters from oil fields have been analyzed for B, Be, Fe, Mn, and Sr by direct emission spectrography using a plasma arc (97).

X-ray fluorescence spectroscopy has had very limited application in natural water and wastewater analysis. The technique is relatively simple and offers a number of advantages over flame photometry and classical spectral emission analysis. While in flame photometry and classical emission spectroscopy the test sample is vaporized and excitation takes place by collisions of individual atoms with each other or with electrons, the sample in the case of X-ray fluorescence undergoes no phase change. The sample is irradiated by a constant X-ray source, and on absorption of primary X-rays, characteristic secondary X-rays are emitted. Measurement is based on the following relationship:

$$I_t = I_p \exp (\mu_m m) \tag{17}$$

where $Ip$ is intensity of the primary radiation, $I_t$ is the intensity of the transmitted radiation, $\mu_m$ is mass absorption coefficient, and $m$ is the mass of irradiated material. The technique is capable of nondestructive analysis for elements in the parts per million range with $\pm 1\%$ accuracy. The test material may be spread in a thin layer prior to analysis (e.g., the sample solution may be applied to a filter paper which is then dried).

X-ray fluorescence has been used for the analysis for As, Se, Hg, Tl, Pb, and Bi in natural waters in concentrations as low as 0.01 ppm (290). The procedure is based on chelating the metals with pyrrolidine thiocarbamate and extracting into chloroform prior to X-ray analysis.

### 3. Activation Analysis

Activation analysis has been applied in many instances for the elemental analysis of natural waters and wastewaters. This technique involves irradiating a test sample (e.g., with thermal neutrons or gamma photons) to induce nuclear transformation of the elements under investigation; these then decay, each with a characteristic half-life. The induced activity is related to the quantity of reactive nuclide by the equation

$$A = \sigma f(W\phi/M)(1 - e^{-\lambda t})(6.2 \times 10^{23}) \tag{18}$$

where $A$ is the activity (disintegrations/sec,) after irradiation; $f$ is the flux of particles used in the irradiation (number/$cm^2$/sec); $\sigma$ is the activation cross section for the nuclear reaction concerned ($cm^2$); $W$ is the weight of the element irradiated (g); $\phi$ is the fractional abundance of the particular isotope of the element concerned; $M$ is the atomic weight of that element; $\lambda$ is the decay constant of the induced radionuclide ($sec^{-1}$), and; $t$ is the irradiation time (sec).

Activation analysis is a technique which offers extreme sensitivity; sensitivities of the order of $10^{-12}$ g are obtainable with neutron fluxes of $10^{12}$ $sec^{-1}$ $cm^{-2}$ (75,240). Applications of activation analysis to natural waters and wastewaters generally require radiochemical separation of the sample and comparison of the activities of the unknown sample and of a known mass of standard treated under identical conditions.

A typical example of procedural format for activation analysis is as follows:

*1.* The water sample may be irradiated in the liquid state or it may be converted to a solid state (by precipitation or freeze-drying) prior to irradiation. This preliminary concentration and separation step is used with very dilute sample solutions or to eliminate interferences.

*2.* Weighed quantities of sample and standard are irradiated in suitable containers for a time sufficient to induce adequate radioactivity in the element to be determined.

*3* A known weight of the element being determined is added to both sample and standard as carrier. Sample and standard are then treated in such a way that the carrier and the element in the sample solution are in the same chemical form.

*4.* Chemical separation is carried out to isolate the element (or suitable compound) from all other nuclides. The chemical "yield" is determined by conventional methods. This makes it unnecessary for the chemical separation to be quantitative, since the final measured activity will be corrected for losses, using the chemical yield figure.

*5.* Measurement and comparison of the radioactivity of sample and standard are made under identical counting conditions, and, when necessary, corrections are applied for self-absorption dead-time losses, decay, etc.

*6.* The radiochemical purity of the separated compound may be checked by determining: (a) decay rate and half-life; (b) energies of emitted radiation, by measuring the activity through different thicknesses of aluminum; and (c) gamma energy by a scintillation counter or solid--state gamma spectrometer.

Depending on the type of sample and purpose of analysis, the use of a carrier and the determination of the chemical yield may be dispensed with.

Characterization of elements in a given water sample may be done by identifying type, energy, and half-life of emission. The concentration of a given element in a sample is determined by quantizing its character-istic emission, i.e.,

$$\frac{\text{Mass of element X in sample}}{\text{Mass of element X in standard}} = \frac{\text{total activity from element X in sample}}{\text{total activity from element X in standard}}$$

Some of the factors governing quantitative measurements are (a) the concentration of stable elements, (b) activation probability, (c) neutron flux, (d) relative abundance of stable isotopes, and (e) time of irradia-tion. A wide range of separation techniques have been associated with activation analysis procedures, some of which have been automated effectively (216). Considerable improvement in the resolution of indi-vidual peaks in spectra has been accomplished by replacement of sodium iodide detectors with germanium detectors (365) for gamma ray scintilla-tion spectrometry.

Detailed procedures for activation analysis of natural waters and wastewaters have been reported (15,62). Generally, the accuracy and precision of these procedures are about ±10%. Principal errors in analy-sis are due primarily to self-shielding, unequal flux at the sample and standard positions, and inaccurate counting procedures and counting statistics. A comparison of the sensitivity of activation analysis with several other analytical methods (304) was given by Meinke. Some of the main advantages of activation analysis are: (a) its very high sensi-tivity, (b) the rapidity of analysis, and (c) the nondestructive nature of the test.

The majority of the studies reported describe procedures for chemical treatment of water and wastewater samples for analysis similar to those discussed earlier (62,257,464,431). One technique which involves the ad-dition of a carrier salt (high purity $Na_2CO_3$, freeze-drying, irradiation) of the recovered residue, and the separation and counting of the desired radioactive nuclides has been reported effective for the analysis of stream and lake waters (464).

The problem of identifying the components of a complex spectrum has led to the use of computers to facilitate interpretation of available data (340). Computers may also be used to determine optimum conditions for irradiation of particular samples (216).

## 4. Electrochemical Analysis

Electrochemical methods are often well suited for metal ion analysis in natural waters and wastewaters. A variety of electrode systems and electrochemical techniques have been used routinely for *in situ* analysis and continuous monitoring of waste effluents.

For purposes of this discussion, electrochemical methods are conveniently classified as being either based on the passage of a faradaic current, e.g., classical polarography, or based on electrode equilibrium, e.g., potentiometry.

Classical polarography has been used widely for water analysis since its development (279). Polarographic measurements are based on determining the time-averaged currents of the dropping mercury electrode under diffusion conditions. The response is described approximately by the Ilkovic equation,

$$i_d = (605nD^{1/2}m^{2/3}t^{1/6})C \qquad (19)$$

where $i_d$ is the average diffusion current ($\mu$A) ; $t$ is the drop-time (sec) ; $m$ is the mass rate of flow of mercury (mg/sec); $D$ is the diffusion coefficient of the electroactive species (cm$^2$/sec) ; $C$ is the concentration of electroactive species (mmoles/liter) ; and $n$ is the number of electrons per molecule involved in the electrode reaction. For a typical case in which $m = 2$ mg/sec, $t = 4$ sec, and $D = 10^{-5}$ cm$^2$/sec, the electrode response will be $i_d/C = 3.82$ $\mu$amp/mmole/liter.

As a result of the capacitance current used in charging the double layer, the sensitivity of classical polarography with the dropping mercury electrode is limited to approximately $10^{-5}$ $M$. However, by means of preconcentration techniques it may be possible to extend the sensitivity range significantly. Copper, bismuth, lead, cadmium, and zinc have been measured in the range of 0.01 mg/liter after extraction with dithizone and carbon tetrachloride (193). Preconcentration by ion exchange, freeze-drying, evaporation, or electrodialysis may be used.

A significant problem in the application of classical polarography to industrial wastewater analysis is the interference produced by electroactive and surface active impurities. Such impurities, frequently present in wastewaters, may interfere with electrode reaction processes and cause a suppression and/or a shift of the polarographic wave (285).

Modifications of polarographic techniques, such as "differential polarography" and "derivative polarography," may be used to increase the sensitivity and minimize the effect of interferences (407). Pulse polarography has the advantage of extending the sensitivity of determination to approximately $10^{-8}$ $M$. The technique is based on the application of short potential pulses of 50 msec on either a constant or gradually increasing background voltage. Following application of the pulse, current measurements are usually made after the spike of charging current has decayed. The limiting current in pulse polarography is larger than in classical polarography. The diffusion current equation for pulse polarography is given by eq. (20). Derivative pulse polarography, which

$$i_d = (nFA(D/\pi t)^{0.5})C \tag{20}$$

is based on superimposing the voltage pulse upon a slowly changing potential (about 1 mV/sec) and recording the difference in current between successive drops versus the potential, is even more sensitive than pulse polarography (407).

Cathode ray polarography or oscillographic polarography has been used for analysis of natural waters and wastewaters, with a sensitivity of $10^{-7}$ $M$ being reported (196,359,499). This technique involves the use of a cathode ray oscilloscope to measure the current–potential curves of applied (saw-tooth) potential rapid sweeps during the lifetime of a single mercury drop. Multiple sweep techniques are also applicable. The peak current ($i_p$) in the resulting polarogram is related to the concentration of the electroactive species for a reversible reaction in accordance with the following relationship:

$$i_p = (k_n^{3/2}m^{2/3}t^{2/3}D^{1/2}v^{1/2})C \tag{21}$$

where $v = dE/dt$ is the voltage sweep (about 6000 V/min).

Oscillographic polarography has the advantage of: (a) relatively high sensitivity, (b) high resolution, and (c) rapidity of analysis. Traces of Cu, Pb, Zn, and Mn can be determined at 0.05 mg/ml level in natural waters by this technique (499).

One of the most interesting electrochemical approaches to metal analysis in trace quantities is anodic stripping voltammetry (414). This technique involves two consecutive steps: (1) the electrolyte separation and concentration of the electroactive species to form a deposit or an amalgam on the working electrode; and (2) the dissolution (stripping) of the deposit. The separation step, best known as the preelectrolysis step, may be done quantitatively or arranged to separate a reproducible fraction of the electroactive species. This can be done by performing the preelectrolysis step under carefully controlled conditions of potential, time of electrolysis, and hydrodynamics of the solution.

The stripping step is usually done in an unstirred solution by applying a potential, either constant or varying linearly with time, of a magnitude sufficient to drive the reverse electrolysis reactions. Quantitative determinations are done by integrating the current–time curves (coulometry at controlled potential) or by evaluating the peak current (chronoamperometry with potential sweep). Several modifications of the separation and stripping steps have been reported (413).

Hanging-drop mercury electrodes of the Gerischer type (166) or Kemula type (232) have been widely used for anodic (or cathodic) stripping analysis. Greater sensitivity has been achieved by use of electrodes which consist of a thin film of mercury on a substrate of platinum, silver, nickel, or carbon (295). Errors due to nonfaradaic capacitance current components can be minimized by proper choice of stripping technique.

The main advantage of stripping voltammetry is its applicability to trace analysis. The technique has been applied for metal analyses in sea water (32), natural waters (277,279), and wastewaters (279).

Electroanalytical methods based on electrode equilibrium include a variety of membrane electrode systems which are applicable for analyses of metals in natural waters and wastewaters. Recent developments in glass electrodes make it possible to use these electrodes to analyze for certain metal ions, particularly sodium and potassium (132). The doping of ordinary glass pH electrodes with $Al_2O_3$ greatly enhances the "alkaline error" or the response of these electrodes to alkalies, and at the same time it reduces their pH response.

It has been found experimentally (132) that for a mixture of two cations, that is, $Na^+$ and $K^+$, the behavior of a modified glass electrode may be described by a modified form of the Nernst equation.

$$E = \text{const.} + RT/F \ln (^\alpha A^+ + K^\alpha B^+) \qquad (22)$$

where $^\alpha Ag^+$ and $^\alpha B^+$ are the activities of $A^+$ and $B^+$ ions, respectively, and $K$ is a selectivity constant which expresses the relative sensitivities of the glass electrode for ions $B^+$ and $A^+$. By varying the composition of the glass in the system $M_2O \cdot Al_2O_3 \cdot SiO_2$ where $M_2O$ is $Li_2O$, $Na_2O$, $K_2O$, $Rb_2O$, or $Cs_2O$, it is possible to vary the selectivity of response of the glass to each of the various alkali metal ions.

Liquid ion exchange has been used in conjunction with potentiometric specific-ion electrode systems. One example is the calcium-selective membrane electrode (392), which consists of the calcium salt of dodecylphosphoric acid dissolved in di-$n$-octylphenyl phosphate. This liquid ion exchanger is immobilized in a porous inert membrane, such as cellulose.

Below a solution concentration of $10^{-6}$ moles $Ca^{+2}$ per liter, the membrane potential is a constant, independent of the calcium ion activity. This has been attributed to the organic calcium salt solubility, which maintains a constant limiting activity of calcium ions at the membrane solution interface (22,392). Otherwise, the liquid ion exchange membrane electrode exhibits behavior similar to that of the glass electrode, which may be expressed in terms of the modified Nernst relationship given in Eq. (22).

Single crystal membrane electrode systems (solid-state membrane electrodes) have recently been applied to water analysis (22,155). The fluoride electrode (155), which is made of lanthanum fluoride crystal membrane doped with a rare earth, which presumably acts so that only fluoride carries current across the membrane, is a good example. This electrode system is highly selective for fluorides, but, at high pH, hydroxide ions constitute a major interference which limits its usefulness in this range. The introduction of anion-selective, precipitate-impregnated membrane electrodes (366,373) has been an important recent development for electrochemical analytical techniques. The membrane is made of a silicone rubber matrix which incorporates precipitated particles of silver halide or barium sulfate. The electrode is relatively insensitive to the nature of the cations. The most selective and sensitive electrode system of this type is the silver iodide membrane electrode, which responds to iodide concentrations as low as $10^{-7}$ $M$ with relatively little interference from common ions (373).

Although the last two membrane electrode systems described above are not sensitive to metal ions, they are included in this section to provide a unified coverage of potentiometric membrane electrodes.

From eq. (22) it is apparent that potentiometric membrane electrode systems are sensitive to the activity of the electroactive species. In order to use such electrode systems for determination of concentrations rather than activities, it is important to consider the effects of ionic strength on the activity coefficient of the electroactive species and the liquid junction potential between the test solution and the reference electrode. To avoid the uncertainty of estimating an activity coefficient, it is useful to determine the effect of an added known amount of species on the potential or adjust the ionic strength of the sample to that of a standard solution. Since the total ionic strength of the solution determines the activity coefficient for a specific ion, the activity coefficient of the ion being analyzed in the test sample will be identical to that in the standard solution. A constant ionic strength can be obtained by using a "swamping electrolyte." This technique, frequently referred to as the "ionic medium" method, has been effectively used to calibrate potentiometric membrane electrode systems for the analysis of natural waters and wastewaters (22,419).

The literature contains conflicting reports regarding the sensitivity limits for the various electroanalytical procedures discussed herein. A helpful discussion of this subject has been given by Laitinen (253); the conclusions of this discussion are in Table X. The sensitivity values given in Table X are defined as the lowest concentrations at which a determination can be made with a relative precision of 10% in the presence of a large concentration of major constituents (253).

A complete analysis of metal ions in natural waters and wastewaters should include definition of oxidation states and characterization of aquometallic or organometallic complexes. Aquometallic complexes in natural waters and wastewaters undergo exchange reactions in which coordinated water molecules are exchanged for certain organic ligands. The pH and the concentration of the organic ligands are important factors affecting such coordination reactions.

Electrochemical methods, in general, are more suited for elucidation of the oxidation state and complexed form in which metal ions exist in a particular sample than are spectrophotometric analyses. Organic polarography, in one or more of the various modes discussed previously, provides a useful method for such determinations. Recent studies have demonstrated the analytical feasibility of thin layer anodic stripping voltammetry for the quantitative estimation of free and bound metals in natural waters (17).

TABLE X
Quantitative Sensitivity Limits of Some Electroanalytical Methods

| Methods | Sensitivity limit, $M$ |
|---|---|
| AC polarography; chronopotentiometry; thin layer coulometry; potentiometry with metal-specific glass or membrane electrodes. | $10^{-4}$ to $10^{-5}$ |
| Classical polarography; coulometry at controlled potential; chronocoulometry; tensammetry; precision null-point potentiometry. | $10^{-5}$ to $10^{-6}$ |
| Test polarography; derivative polarography; square-wave polarography; second harmonic AC polarography; phase-sensitive AC polarography; linear sweep voltammetry; staircase voltammetry; derivative voltammetry; coulostatic analysis; chemical stripping analysis. | $10^{-6}$ to $10^{-7}$ |
| Pulse polarography; rf polarography; coulometric titrations; amperometry with rotating electrodes; conductivity (aqueous). | $10^{-7}$ to $10^{-8}$ |
| Anodic stripping with hanging-drop mercury electrodes. | $10^{-7}$ to $10^{-9}$ |
| Anodic stripping with thin film electrodes or solid electrodes. | $10^{-9}$ to $10^{-10}$ |

Membrane electrodes, either of the voltammetric or the potentiometric type, are usually sensitive only to free or unbound metal ions in a water sample. For example, in one case in which the calcium membrane electrode was used in a sea water sample, only 16% of the total calcium content was detected (458); the authors explained this in terms of the remainder of the calcium being complexed with sulfates and/or carbonates.

Future prospects for metal ion analyses in waters and wastewaters include the use of gas chromatography and NMR techniques. It would appear that more emphasis will be placed on the characterization of organometallics in wastewaters than on elemental analysis per se.

## VII. ANALYSES FOR NONMETAL INORGANIC SPECIES

Discussion in this section is directed to the analysis of nonmetal inorganic species other than nitrates and phosphates, which, because they are frequently classified as biochemical nutrients, have been discussed in the section on organic analysis.

### A. SEPARATION AND CONCENTRATION TECHNIQUES

The techniques discussed for separation and concentration of metal ions are generally applicable also for nonmetal inorganic species. Evaporation, precipitation, ion exchange, solvent extraction, and partial freezing are frequently used, ion exchange being particularly well suited for anion separations. Anion exchange chromatography has been used rather extensively for separations of species found in waters and wastewaters (495). Silicates, for example, can be effectively separated from natural waters by treating with hydrogen fluoride to form fluorosilicates, which are then removed by exchange (500).

Techniques for the chromatographic separation of ortho-, pyro-, tri-, trimeta, tetrameta-, and polyphosphates have been developed (174,492) with strongly basic anion exchange resins. Potassium chloride, in continuously increasing concentration (gradient elution technique), is used for elution. Sulfate, sulfite, thiosulfate, and sulfide ions can be separated by anion exchange chromatography using the gradient elution technique with an elution solution of nitrates (211). A more detailed discussion of the possible applications of anion exchange for such analytical separations has been given by Inczedy (214).

## B. INSTRUMENTAL METHODS

A number of instrumental methods are applicable to the analysis of nonmetallic inorganic species in natural waters and wastewaters. Many such analytical methods involve either the use of potentiometric membrane electrodes or the application of spectrophotometric techniques. New methods of absorption spectrophotometry using solid-phase ion (or ligand) exchange reagents have been devised for analyses for chloride, sulfates, phosphates, and other anions (44,57,188). These tests are based on using various salts of chloranilic acid as selective ion exchange reagents. Typical examples include the analysis for sulfates with barium chloranilate (57), for chlorides with mercuric chloranilate (44), and for phosphates with lanthanum chloranilate, i.e.,

$$BaCh + SO_4{}^{2-} \rightarrow BaSO_4 + Ch^{2-} \qquad (23)$$
$$HgCh + 2Cl^- \rightarrow HgCl_2 + Ch^{2-} \qquad (24)$$
$$La_2Ch_3 + 2PO_4{}^{3-} \rightarrow 2LaPO_4 + 3Ch^{2-} \qquad (25)$$

This technique is subject to interferences which may be reduced by solvent extraction.

Many highly sensitive methods for nonmetallic inorganic species are based on the displacement of a ligand (usually colored) in a metal complex or chelate. The analysis for fluoride ions by displacement of a chelating dye anion from a zirconium complex is a typical example of this technique. The displaced dye differs sufficiently in color from its zirconium chelate to permit determination of the fluoride ion concentration as a function of the change in color of the solution (54).

Numerous indirect spectrophotometric methods have been developed (65). Heteropoly chemistry appears to offer important advances in analyses for phosphates, silicates, and arsenates (104). The reader is referred to recent reviews in *Analytical Chemistry* for an exhaustive survey on the subject matter (65).

Indirect uv spectrophotometry and atomic absorption methods have been developed for phosphates and silicates (210). These techniques are based on the selective extraction of molybdophosphoric acid and molybdosilicic acid followed by ultraviolet molecular absorption spectrophotometry and/or atomic absorption spectrophotometry. The molybdophosphoric and molybdosilicic acids are formed in acidic solution by addition of excess molybdate reagent. Molybdophosphoric acid is extracted with diethyl ether from an aqueous solution which is approximately 1 *M* in hydrochloric acid. After adjusting the hydrochloric acid concentration of the aqueous phase to approximately 2 *M*, the molybdosilicic acid is extracted with 5:1 diethyl ether–pentanol solution. The extracts of molybdophosphoric and molybdosilicic acids are subjected to acidic washings to remove excess molybdate. Each extract is then contacted with a basic buffer solution to strip the heteropoly acid from the organic phase. The molybdate resulting from the decomposition of the heteropoly acid in the basic solution is then determined either by measurement of the absorbance at 230 $m\mu$ using ultraviolet spectrophotometry or by measurement of absorbance at the 313.3 $m\mu$ resonance line of molybdenum by atomic absorption spectrophotometry. The optimum concentration ranges are approximately 0.1–0.4 mg/liter of phosphorus or silicon for indirect ultraviolet spectrophotometry and 0.4–1.2 mg/liter for indirect atomic absorption spectrophotometry.

Electrochemical methods of analysis offer considerable promise for selective and specific measurements in wastewaters. Direct potentiometric techniques using electrodes of the second kind may be used for the analysis of chlorides or sulfides. Silver electrodes, coated with a layer of halides or sulfides, are available commercially for determination of chlorides, bromides, iodides, and sulfides at concentrations corresponding to the solubilities of the respective silver salts, as given by the following expressions:

$$E_{cell} = E^{\circ}_{Ag^+,Ag} - E_{ref} + \frac{RT}{nF} \ln K_{sp} - \frac{RT}{nF} \ln Cl^- \qquad (26)$$

and

$$E_{cell} = E^{\circ}_{Ag^+,Ag} - E_{ref} + \frac{RT}{nF} \ln K_{sp} - \frac{RT}{nF} \ln S^{2-} \qquad (27)$$

where $K_{sp}$ refers to the solubility products of AgCl and $Ag_2S$, respectively, in eqs. (26) and (27). Limitations on the use of such electrode systems are imposed by interferences from other potential-determining ions, by the problem of elimination of liquid junction potentials, and by the difficulty of satisfactorily resolving single ion activity coefficients.

Major developments in electrochemical analysis for anions have occurred in the area of ion-selective electrodes. Such electrode systems are primarily solid state or precipitate ion exchange membrane electrodes. Pungor and co-workers (366) have developed anion selective membrane electrodes by impregnating silicone gum rubber membranes with specific insoluble salts. Electrodes which are selective for $Cl^-$, $Br^-$, $I^-$, $S^{2-}$, $SO_4^{2-}$, and $F^-$ have been reported (367,368). Punger and Havas (367) have reviewed the literature of ion-selective membrane electrodes and discussed the preparation and characterization of the membrane. Response time, memory effects, and detection limits up to $10^{-4}$ $M$ have been discussed. Detailed studies on the sensitivity and selectivity of iodide, bromide, and chloride electrodes has been reported by Rechnitz et al. (373,374). The electrodes were prepared by incorporating AgI, AgBr, and AgCl into silicone rubber matrices. Selectivity ratios for the various electrodes were reported with response times $(t_{\frac{1}{2}})$ of 8, 14, and 20 sec, respectively. The lower limits of response (Eq. (22)) were $10^{-7}$ $M$ $I^-$, $5 \times 10^{-4}$ $M$ $Cl^-$, and $7 \times 10^{-5}$ $M$ $Br^-$.

Electrode systems prepared by incorporating metal oxides and hydroxides (167) or ion exchange resins (366) in polymeric membranes also have been reported. Solid-state ion exchange membrane electrodes for fluoride ion determinations have been reported (155). This type of membrane electrode is made of single crystals of $LaF_3$, $NdF_3$, and $PrF_3$ and is reported to be sensitive to fluoride ion concentrations as low as $10^{-5}$ M.

Anion-selective membrane electrodes can be used for direct potentiometry or for potentiometric titrations. For example, Lingane (265) used a commercially available fluoride electrode for the potentiometric titrations of $Th^{4+}$, $La^{3+}$, and $Ca^{2+}$ ions. Best results were obtained with $La(NO_3)_3$. The equivalence point potential was determined with $\pm 2$ mV, and an accuracy better than 0.1% noted in neutral, unbuffered solutions.

Cathodic stripping voltammetry can be also used for trace analysis of halides (414). Brainina and co-workers (51) have compared the limits of sensitivities of various anions using mercury electrodes and have given concentration limits of $5 \times 10^{-6}$ $M$ for $Cl^-$, $1 \times 10^{-6}$ $M$ for $Br^-$, and $5 \times 10^{-6}$ $M$ for $I^-$.

## C. SULFUR COMPOUNDS

Sulfides commonly occur in a variety of waste effluents (e.g., septic sewage, oil refinery wastes, tannery wastes, viscose rayon wastes, etc.). The presence of sulfides in wastewaters is usually indicated by the characteristic odor of hydrogen sulfide. The acidity constants for the diprotic acid $H_2S$ are $K_1 = 1.0 \times 10^{-7}$ and $K_2 = 1.2 \times 10^{-13}$.

Detection of free sulfides in wastewaters is relatively straightforward; a small sample volume is placed in a 150-ml glass-stoppered conical flask and slightly moistened lead acetate paper is suspended between the stopper and the neck. On acidification of the sample a brown stain on the paper indicates the presence of sulfides. As low as 0.01 mg/liter $H_2S$ in a 50-ml sample can be detected in this manner.

Total sulfides may be accurately determined in the range from 0.1 to 20 mg/liter $H_2S$ by standard molecular absorption spectrophotometry (360). The test is based on the fact that hydrogen sulfide and sulfides react with $p$-aminodimethylaniline hydrochloride in the presence of sufficient hydrochloric acid and an oxidizing agent (ferric chloride) to produce methylene blue dye. The test is sensitive to free sulfides, as well as sulfides bound by iron, manganese, and lead. Sulfides of copper and mercury are too insoluble to react. Sulfites and thiosulfates interfere, but by increasing the amount of ferric chloride and lengthening the time of reaction, up to 50 mg/liter of these compounds may be tolerated.

Sulfides in industrial waste effluents may be precipitated by adding zinc or cadmium acetate or ammoniacal zinc chloride. The precipitated sulfide is then added to excess acidified standard iodine solution which is back-titrated with standard thiosulfate (233). Several modifications of iodometric determinations of sulfides in wastewaters have been reported (208,364).

The methods of analysis for sulfides described above give total sulfides, i.e., free and bound sulfides, polysulfides, and sulfanes. A specific test for only free sulfides can be accomplished using specific ion exchange membrane electrodes (283).

Specific ion electrode potential measurements of $S^{2-}$ can be related to the concentrations of $H_2S$, $HS^-$, and $S_T$ (analytical concentration of free sulfides) by the following equilibrium relationship:

$$\log [S_T] = \log [S^{2-}] + \log \left( \frac{[H^+]^2}{K_1K_2} + \frac{H^+}{K_1} + 1 \right) \tag{28}$$

$$\log [H_2S] = \log [S^{2-}] + \log \frac{[H^+]^2}{K_1K_2} \tag{29}$$

$$\log [HS^-] = \log [S^{2-}] + \log \frac{[H^+]}{K_1} \tag{30}$$

where $K_1$ and $K_2$ are the acidity constants for the weak acid $H_2S$.

Sulfites are commonly found in the wastewaters from pulp and paper mills, in which sulfites are used for the preparation of cellulose from wood. In the absence of thiosulfates, sulfites can be detected by heating an acidified water sample and identifying the evolved sulfur dioxide by the blue coloration produced when subjected to a piece of filter paper moistened with a mixture of potassium iodate solution and starch solution. Sulfites also may be detected in waste effluents by decolorization of triphenylmethane dyes by neutral sulfite solutions.

The iodometric titration of sulfites is done after sulfides are precipitated and filtered as zinc sulfides. This test is subject to interferences from organic matter and other reduced compounds, such as ferrous iron, in the test solution.

Sulfate determination in wastewaters is not called for except in connection with problems associated with corrosion of concrete pipes. Sulfates may be determined gravimetrically or turbidimetrically when precipitated as barium sulfate (20). Sulfites, sulfides, silica, and other insoluble solids may interefere with the gravimetric procedure. Turbidimetric measurements of $BaSO_4$ may be improved by using glycerin and sodium chloride to stabilize the suspension.

Titrimetric analysis of sulfates may be done by first precipitating the sparingly soluble benzidine sulfate, followed by titration of the washed precipitate with standard sodium hydroxide using phenolphthalein as an indicator (16). Interferences by phosphates can be minimized by using a colorimetric technique. The precipitated benzidine sulfate is washed with an alcohol–ether mixture to remove excess benzidine and is dissolved in 1% sodium borate, and the liberated benzidine is allowed to react with 1:2 naphthoquinone–4 sulfonate to give a red color which is determined photometrically at 490 $m\mu$ (236).

Sulfate ions can be also titrated in an alcoholic solution under controlled acid conditions with a standard barium chloride solution using thorin as an indicator. This test can be extended by use of ion exchange separation and concentration techniques. A number of anions and cations may cause interferences (e.g., potassium, iron, aluminum, phosphates, fluorides, and nitrates). Most metallic ions also interfere by forming colored complexes with the thorin indicator, especially in the alcohol–water solution. It is sometimes necessary to use ion exchange separation to remove interferences prior to sulfate titration (159).

## D. CYANIDES, THIOCYANATES, AND CYANATES

Cyanides may be found in significant quantities in industrial wastewaters from electroplating, steel, coke ovens, gold mining, and metal finishing industries. Cyanides are extremely toxic, even in low concentrations. In view of their toxicity, they can be detected easily by fish-kill incidence in concentrations as low as 0.03 mg/liter HCN.

Hydrogen cyanide is a volatile weak acid ($K_a = 4.8 \times 10^{-10}$). Both HCN and the conjugate base, $CN^-$, are called "free cyanides." Stable cyanide salts and metal cyanide complexes, such as $K_4Fe(CN)_6$ and $K_3Co(CN)_6$, also may be found in waste effluents. Total cyanide determinations include both free cyanides and complexed cyanides. The test for total cyanides (20) usually involves breaking down complex metal cyanides by distillation with (a) hydrochloric acid solution of cuprous chloride, (b) tartaric acid, (c) mercuric chloride, magnesium chloride, and sulfuric acid (410) (Serfass method); and (d) phosphoric acid in the presence of citric acid and ethylenediaminetetraacetic acid (248).

A simple procedure for detecting cyanides in a wastewater is based on treating a strip of filter paper with a drop of 10% ferrous sulfate and a drop of 10% caustic soda and suspending the paper strip over the acidified sample solution in a glass-stoppered flask for about 10 min. The liberated HCN reacts to form ferrocyanide on the paper, which, when immersed in hot dilute hydrochloric acid, yields the blue to blue–green stain of prussian blue dye. This test is sensitive to about 0.4 mg/liter of HCN.

Traces of cyanides can be determined by the Aldridge test, frequently known as the bromin–pyridine–benzidine method (12). The procedure utilizes the König synthesis, which is based on converting cyanide to cyanogen bromide (or chloride) and the reaction with pyridine and an aromatic amine (benzidine) to give an intense characteristic color. The di-anil derivative formed (orange to red) may be determined spectrophotometrically. Thiocyante will give the same reaction as cyanides with this test. Interference due to thiocyanide may be minimized by extracting cyanides from acid solution with isopropyl ether, which can be recovered by extraction in an aqueous sodium hydroxide solution (248). An alternate procedure (236) is based on cyanide conversion to cyanogen chloride by chloramine-T, which is allowed to react with pyridine containing 1-phenyl-3-methyl-5-pyrozolone to give a blue color which is determined spectrophotometrically (20).

A very sensitive and accurate method for the analysis of cyanides is based on the interaction of "ferroin" with cyanides at pH 9.2–9.7 (404). The violet complex [dicyano-bis-(1:10-phenanthroline)-iron] formed is extracted with chloroform and determined spectrophotometrically. This procedure is claimed to be subject to fewer interferences than any of the other methods discussed previously.

One of the oldest procedures for the analysis of cyanides in waste effluents is the Liebig test (20). This is based on the titration of cyanides with silver nitrate to form a soluble argentocyanide complex, $[Ag(CN)_2]^-$. After all cyanides have reacted, excess silver nitrate will cause turbidity due to precipitation of silver cyanide. The end point is characterized by the first appearance of turbidity, which in most cases is difficult to detect by eye. A colorimetric indicator, p-dimethylaminobenzylidine-rhodanine in acetone, can be used to aid in detection of the titration end point. In alkaline solution (pH 10–11) the end point is characterized by a change of color from yellow to salmon-pink. Dithizone may be also used for this titration (404).

Thiocyanates and cyanates are frequently present in plating and gasworks waste effluents. The cyanates may be easily detected by their interaction with ferric chloride in hydrochloric acid solution to give a characteristic wine-red color. This test is very approximate in nature and is subject to various interferences, such as those caused by mercury salts, fluorides, and organic hydroxy acids (6).

Thiocyanates, if present in relatively large quantities, can be estimated by precipitation as cuprous thiocyanate, which is either titrated with standard potassium iodate in acidic solution in the presence of chloroform or is decomposed with caustic soda, acidified with nitric acid, and titrated with silver nitrate using alum as indicator (233).

Thiocyanates may be determined colorimetrically in trace quantities by the formation of the copper pyridine thiocyanate complex $[Cu(C_5H_5N)_2 (CNS)_2]$, which can be extracted in chloroform to yield a yellow solution (249). The test is sensitive to 0.5 mg/liter $CNS^-$ and is subject to interferences from cyanides. Cyanides may be removed by acidifying the solution with acetic acid and stripping HCN by aeration.

Cyanates may be determined by hydrolysis of the cyanates to ammonia by boiling with acid and estimation of ammonia so formed by means of Nessler's reagent (162). Any ammonia originally present in the sample should be removed.

## E. CHLORIDES

Chlorides are present in large quantities in certain pickle liquors, in spent regenerant solutions from softening plants, oil well waters, drainage from irrigation waters, and other waste effluents.

The terms "salinity" and "chlorinity" are frequently used to express chloride content in marine and estuarine waters. "Salinity" is defined as the weight in grams of dissolved solids (dried to constant weight at 480°C to remove organic matter and to convert carbonates to oxides) in 1000 g of sea water. "Chlorinity," on the other hand, is defined as the weight of halides, as measured by reaction with silver nitrate and computed on the basis of all halides being represented as chlorides. Both salinity and chlorinity are commonly expressed in parts per milli and are related in sea water as follows:

$$S\%_0 = 0.03 + 1.805 \ Cl\%_0 \tag{31}$$

where $S\%_0$ and $Cl\%_0$ refer to parts per milli salinity and chlorinity, respectively.

One procedure for chlorides in wastewaters is by titration with silver nitrate. In this method, which is commonly referred to as the Mohr method, the end point is detected by the formation of reddish silver chromate by the slightest excess of silver. The test is subject to interferences from substances such as sulfides, thiocyanides, phosphates, cyanides, sulfites, acids, alkalies, and any anion which may form a sparingly soluble silver salt. Some of these interferences can be eliminated by adjustment of pH or by potentiometric titration with a silver wire indicator electrode and a glass electrode as a reference (375). Volatile sulfur compounds which may be present in wastewaters from oil refineries and synthetic rubber manufacturing may be removed by evaporation. Sulfides and thiocyanates may be destroyed by addition of hydrogen peroxide and heating.

If the test solution contains large quantities of phosphates, the Volhard procedure is recommended (20). This involves addition of an excess of silver nitrate to a sample which has been acidified with dilute nitric acid, followed by coagulation of the silver chloride by shaking with nitrobenzene, and, finally, back-titration with standard thiocyanate using gerric alum as indicator.

Titration of chloride with mercuric nitrate offers a more sensitive procedure (20). The test is based on titration with mercuric nitrate to form a slightly dissociated mercuric chloride complex in an acid solution (pH $\approx$ 3.0). The end point is detected by the mixed indicator diphenylcarbazone and bromophenol blue, which turns blue–violet upon addition of a slight excess of mercuric ion.

## F. FLUORIDES

Fluorides occur in the wastewaters of industries involved in the production of aluminum from bauxite, in the production of phosphatic fertilizers, in oil field effluents, in nuclear power plant effluents, and in wastewaters from the scrubbing of flue gases and from the etching of glass.

Fluorides may be titrated with standard thorium nitrate in a solution buffered at pH $\approx$ 3.0 (67). Sodium alizarin sulfonate, solochrome brilliant blue, and chrome azurol S have been used as end point indicators (67, 314). Interferences from sulfate, phosphate, and carbonate can be minimized by sample pretreatment with barium chloride.

The most popular method of fluoride analysis is based on the bleaching action of fluorides on the reddish lake formed by zirconium oxychloride and sodium alizarin sulfonate (254). Visual color comparison is usually done with the zirconium alizarin test (20). Both color intensity and hue vary with fluoride concentration. A more stable colored complex is formed with fluorides and zirconium-Eriochrome cyanine R (20). The color intensity of the complex is diminished by addition of F⁻ and is measured spectrophotometrically at 540 $m\mu$.

The preceding colorimetric procedures are subject to numerous interferences commonly found in industrial wastewaters. It is therefore recommended that fluorides be separated by distillation or ion exchange (20) prior to analysis by the above procedures.

Probably the most direct and simple method for the analysis of fluorides is based on the use of potentiometric ion selective membrane electrode systems. Lanthanum fluoride membrane electrodes have been successfully used for direct potentiometric measurements of fluorides in water supplies (155b) and industrial wastewaters (155a). The membrane electrode has also been used for precise end point detection in the titration of fluoride with lanthanum, calcium, and other fluoride-complexing agents (265).

Hydroxide ions at high concentrations interfere with the lanthanum fluoride electrode sensitivity, which limits its application at high pH values. The electrode sensitivity is estimated to be about $10^{-6}$ MF⁻ or 0.02 ppm.

## VIII. DISSOLVED GASES

The present discussion of analyses for dissolved gases in waste effluent is concerned only with elemental gases, such as molecular nitrogen and oxygen. Analyses for important volatile inorganic weak acids and bases, such as $H_2S$, $CO_2$, and $NH_3$, have been discussed in earlier sections of this chapter.

### A. SEPARATION AND CONCENTRATION TECHNIQUES

Dissolved gases in waste effluents usually may be separated rather readily by vacuum degasification or by one or more of various stripping techniques. Stripping is essentially a gas–liquid extraction procedure in which an inert carrier gas is bubbled through a sample to carry off the dissolved gases for further separation, concentration, or detection. Gas transfer efficiency in such systems is dependent on the gas–liquid interfacial area and on the degree of mixing.

Gas exchange separation can be carried out as either a batch or a continuous flow process. In one design, a continuous mixed stream of sample and carrier gas (nitrogen or hydrogen) is forced through an aspirator nozzle under 50 lb of pressure (453). In another design, the dissolved gases are stripped from the test solution by means of multiple spinning disks rotating at high speed (446,503). Detection of the stripped gases in the stream of carrier gas may be done by measurement of paramagnetic susceptibility, thermal conductance, etc. (284).

The gas stripped from a wastewater sample may be separated into its various components by gas chromatography (503). Several modifications of this technique have been reported (442,446,447,453,503). By choosing appropriate detectors, it is usually possible to analyze simultaneously for almost all gases of interest in a water or wastewater sample.

## B. DISSOLVED OXYGEN

A detailed discussion of analytical methods for dissolved oxygen will serve to exemplify various techniques applicable to the analysis of most dissolved gases of interest. The analysis of dissolved oxygen in industrial wastewater has always been considered a highly significant test with respect to characterization of the physicochemical and biochemical characteristics of a waste effluent and its effect on a receiving water.

*In situ* analysis is probably the most effective way to analyze for dissolved oxygen. Certain precautions should be taken in cases where water samples are collected and stored for subsequent analysis. The sample must not remain in contact with air nor be agitated; either condition will cause a change in dissolved gas levels. Samples from any depth or from waters under pressure require special procedures to eliminate the effects of changes in pressure and temperature on sampling and storage. Detailed description of procedures and equipment for proper sampling of waters under pressure, as well as waters at atmospheric pressure, are available in the literature (20).

The time lag between sampling and analysis is of great significance. The longer the lag, the greater the chance that the oxygen content will change because of chemical or biological activity in the test solution.

The oldest and one of the most popular methods for the analysis of dissolved oxygen is the Winkler test (505a). Originally reported about 75 years ago, the Winkler procedure possesses most attributes of basic soundness and sensitivity. Improved by variations in equipment and techniques and aided by modern instrumentation, this test is still the basis for the majority of titrimetric procedures for dissolved oxygen. The test is based on the quantitative oxidation of manganese (II) to manganese(IV) under alkaline conditions. This is followed by the oxidation of iodide by the manganese(IV) under acid conditions. The iodine so released is then titrated with thiosulfate in the presence of a starch indicator. The reported precision of the standard Winkler test is ±0.1 mg/liter of dissolved oxygen (29).

In applying the Winkler test for oxygen determinations in wastewaters, full consideration must be made of the interfering effects of oxidizing or reducing materials in the sample. The presence of certain oxidizing agents liberates iodine from iodide (positive interference), and the presence of certain reducing agents reduces iodine to iodide (negative interference). Reducing compounds may also inhibit the oxidation of the manganous ion. Certain organic compounds have been found to interfere with the Winkler test in a different way. Surface active agents for example, have been reported to hinder the settling of manganic oxide floc, thus partially obscuring the end point of the final titration with thiosulfate (284).

Several modifications of the Winkler test have been devised to minimize the effect of interferences found in different wastewaters. The Alsterberger modification (19) was developed to eliminate the effect of nitrites, which interfere with the iodometric titration according to eqs. (32) and (33).

$$2NO_2^- + 2I^- + 4H^+ \rightarrow I_2 + N_2O_2 + 2H_2O \qquad (32)$$

$$N_2O_2 + \tfrac{1}{2}O_2 + H_2O \rightarrow 2NO_2^- + 2H^+ \qquad (33)$$

The nitrites formed in the reaction depicted in eq. (33) oxidize more iodide ions to free iodine; thus a cyclic reaction yielding erroneously high results is established. Also, the presence of nitrites in the test solution makes it impossible to obtain a permanent end point. As soon as the blue color of the starch–iodine complex disappears, the nitrites react with more iodide to form iodine, and the blue color of the indicator appears again.

The Alsterberg modification utilizes sodium azide, which reduces the nitrites according to eqs. (34) and (35).

$$NaN_3 + H^+ \rightarrow HN_3 + Na^+ \qquad (34)$$

$$HN_3 + NO_2^- + H^+ \rightarrow N_2 + N_2O + H_2O \qquad (35)$$

The effect of a wide variety of reducing agents may be overcome by the Rideal-Stewart modification (378) of treating the sample with potassium permanganate solution under acid conditions. However, the difficulty in manipulation of the Rideal-Stewart modification often results in low accuracy and precision.

The alkaline hypochlorite modification (455) was designed to overcome interferences caused by sulfur compounds. Wastes from the sulfite pulp industry, for example, usually contain appreciable quantities of sulfites, thiosulfates, polythionates, etc. This procedure involves pretreatment of the sample with alkaline hypochlorite to oxidize the sulfur compounds to sulfates and sulfur. Excess hypochlorite is destroyed by potassium iodide, and the iodine is then titrated by sodium thiosulfate. Again, this modification is difficult to perform, and the results obtained are of low accuracy (284).

Suspended solids in water samples may consume certain quantities of iodine during the Winkler test (284). This interference may be removed by flocculation with alum and ammonium hydroxide and settling of the solids prior to conducting the Winkler test.

For samples with biological activity (e.g., activated sludge or fermentation or food processing waste effluents) the addition of copper sulfate to the sample prior to the test coagulates the biological forms. Sulfamic acid is also added to inhibit biological activity. After allowing the floc to settle, an aliquot of supernatant liquor is siphoned and analyzed for dissolved oxygen by the standard Winkler test. The copper sulfate and the sulfamic acid are combined into a single solution in practice. This modification commonly suffers from relatively low precision (284).

The use of cerous salts in place of manganous salts has been reported to result in less interference by organic compounds (258). Attempts to minimize interferences by oxidizing agents have included sample treatment with sodium hydroxide and ferrous ammonium sulfate (136). Oxygen in the test solution oxidizes the ferrous iron to ferric iron, which is then titrated with ascorbic acid using 4-amino-4'-methoxydipenylamine as an oxidation–reduction indicator.

Interferences also may be minimized by the iodine-difference modification. This method (356) involves the addition of a small quantity of iodine solution to two portions of the water, one of which serves as a control and the other of which is analyzed by the standard Winkler method. Reducing substances in the wastewater react with the iodine in both the sample and the control; therefore, the difference between the standard titration of the sample and that on the control gives an accurate value for the dissolved oxygen in the sample. It is important to maintain the same experimental conditions and time of reaction for both the sample and control after addition of the iodine solution.

Another procedure has been used to compensate for interferences present in wastewaters, as well as those which might appear in the reagents (258). Two water samples, A and B, of equal volume are used. Sample A is analyzed by the standard Winkler test. The order of addition of reagents to sample B, the blank, is reversed, i.e., KOH and KI are added first, followed by $H_2SO_4$ and $MnSO_4$. In the case of sample B, dissolved oxygen is not first "fixed" by the manganous salt. Because the blank is initially made alkaline, allowance is made for substances that interfere specifically in an alkaline medium. Interfering substances in the test solution, as well as those in the reagents, react in both samples. Accordingly, the difference between the sample and the blank (A minus B) represents the oxygen in the sample plus the oxygen in the reagents.

If reducing substances such as sulfite are present in greater amounts than oxidants, it is possible to have a "negative" blank. In this case identical quantities of iodine or iodate are added to both sample and blank to provide an excess of iodine after acidification. The reverse-addition method is commonly used, but it does not correct for interferences caused by ferrous iron. This leads to low results due to the formation of ferrous hydroxide, which reacts with oxygen in the blank. The ferric ions formed do not produce an equivalent amount of iodine upon acidification.

The Winkler test may be modified to titrate dissolved oxygen chlorometrically instead of iodometrically. The test is called the "$o$-tolidine method" (310). After the dissolved oxygen has oxidized the divalent manganese in the conventional Winkler test, the solution then may be acidified with chlorine-free concentrated hydrochloric acid. This reaction results in the liberation of an equivalent quantity of chlorine according to reaction (36).

$$MnO_2 \cdot H_2O + 4H^+ + 4Cl^- \rightarrow Mn^{2+} + 3H_2O + 2Cl_2 \tag{36}$$

The free chlorine then reacts with $ortho$-tolidine to form the characteristic yellow-colored haloquinone chloride. The intensity of the color of the haloquinone chloride is directly proportional to the oxygen content of the sample and is measured by comparison with standards or colored slides or by the use of absorption spectrometry. Recent studies have revealed that the intermediate chlorine is not necessary for oxidation of the $ortho$-tolidine. Colloidal manganese(IV), produced by air oxidation of manganese(II), is capable of reacting directly, even in neutral solution, with $ortho$-tolidine.

The precision and accuracy of titrimetric procedures for dissolved oxygen may be considerably improved by using better end-point detection techniques. Potentiometric detection of the iodometric end point improves sensitivity to about ±0.001 mg/liter of iodine (284). "Dead stop" end-point detection (an amperometric technique) offers an extremely sensitive, as well as accurate, measurement (284). The procedure is quite simple and utilizes two smooth platinum electrodes with a small potential difference (from 15 to 400 mV, depending on the sensitivity required). Diffusion current is measured during the course of the titration. No attempt is made to control the potential of either electrode; only the potential difference is controlled. The end point is indicated by discontinuation of current flow in the cell. As long as free iodine remains in the solution, the chief electrode reaction under the influence of the applied voltage is the oxidation of iodide to iodine at the anode and the reverse process at the cathode. At the end point, when all free iodine has been removed, the iodine to iodide reaction can no longer occur and the cell current comes to a "dead stop." Since the thiosulfate/tetrathionate reaction is highly irreversible and proceeds at only a minute rate under the influence of the applied voltage, no detectable current is observed at and beyond the end point. Ordinarily, the end point is so easily detected that there is no need for a graphical estimation of its position. By using sensitive current-measuring devices the end point can be established to an accuracy of $\pm 0.01$ $\mu$g iodine in a 100-ml sample, or 1 part in 10 billion.

Coulometric titration of dissolved oxygen by *in situ* electrochemical generation of iodine has been used with considerable success (203). The procedure consists of the successive additions of standard solutions of $MnSO_4$, $KOH + KI$, $H_2SO_4$, and an excess of $Na_2S_2O_3$ to the test solution. The iodine formed electrolytically reacts with the residual thiosulfate in solution. The electrolytic current is held constant by varying the potential, and the equivalence point is conveniently detected by the dead-stop end-point method.

The procedure is very accurate, to within 0.02 g/liter. It has a distinct advantage in that, because the titrant is generated in solution, errors caused in conventional titrations by contact with air are eliminated.

Direct colorimetric methods of analysis for dissolved oxygen are based on the interaction of molecular oxygen with an oxidation–reduction indicator to give a color change. One of the most commonly used indicators for the detection of oxygen in solution is methylene blue; others are indigo carmine and safranin T.

In the presence of dissolved oxygen a reduced methylene blue solution exhibits a blue color, and in the absence of dissolved oxygen it is colorless. This indicator has been used in the relative stability test for sewage effluents (29). Quantitative colorimetric determinations of dissolved oxygen can be made with indigo carmine dyes. Indigo carmine in the reduced state reacts with oxygen to give a color change through orange, red, purple, blue, and finally a blue–green in the completely oxidized form. Colorimetric procedures are subject in general to a variety of interferences which limit their applicability to industrial wastewaters.

A radiometric procedure for monitoring dissolved oxygen is based on the quantitative oxidation of radioactive thallium-204 by oxygen in the test solution (377). Thallium-204 is primarily a beta emitter with a half-life of 3.6 years; therefore, decay over several months does not greatly reduce the sensitivity of the technique. The apparatus consists of a column of radioactive thallium electrodeposited on copper turnings, and two flow-type Geiger-Mueller counters.

The technique involves passing the test solution by one of the Geiger-Mueller counters to detect background beta activity, then through the column where reaction (37) occurs.

$$4Tl + O_2 + 2H_2O \rightarrow 4Tl^+ + 4OH^- \tag{37}$$

The radioactive thallium in the effluent from the column is detected by the second Geiger-Mueller counter. One milligram of oxygen liberates $25.6 \times 10^{-3}$g of $^{204}$Tl. The counting rate is directly proportional to the oxygen concentration in the test solution.

The sensitivity of the test using a column with a specific activity of 2.04 mCi per gram of thallium is about 0.2 mg/liter. That is to say, a test solution containing 0.2 mg/liter produces a $^{204}$Tl counting rate equal to the background counting rate of the detector. As a rule, because of the randomness of radioactive disintegrations, the precision of this method is ±2%. It is important to note that oxidizing agents and changes in the pH of the test solution may interfere with the test.

A few coulometric methods of analysis for dissolved oxygen have been reported. For purposes or orientation, it is helpful to note that air-saturated water at 25°C and under 750 mm Hg air pressure contains 8.18 mg of dissolved oxygen per liter. This in terms of coulometric response is 0.1083 amp.sec/g, a rather large quantity. Two coulometric procedures are discussed here (42,284). In one method a deoxygenated solution of chromic ions is added to the water sample. During the test, chromous ions are generated electrolytically and are then reoxidized by the oxygen in solution. The apparatus immediately detects any excess of chromous ions, marking the end of the titration. This method is sensitive to 0.3 mg dissolved oxygen per liter and has an accuracy of ±2%.

Another coulometric method is based on the interaction between oxygen and an ammonia–copper complex, $Cu(NH_3)_2^+$. This is followed by the reduction of the oxidized ammonia–copper complex on a platinum cathode. The amount of current used is equivalent to the oxygen concentration in the test solution.

Use of a constant-potential derivative coulometric system was reported recently (125). The working electrode was composed of tiny metal spheres packed in a ceramic tube. No field experience has been reported with this system, however.

Voltammetric analyses for dissolved oxygen in wastewaters have been carried out with various degrees of success using rotating platinum electrodes and dropping mercury electrodes. The main difficulty in using such electrode systems in industrial waste effluents is the presence of surface active and electroactive interferences which frequently cause "electrode poisoning." A detailed discussion of the effects of surface active agents on the polarographic oxygen determination is available in the literature (285).

Various modifications of the dropping mercury electrode system have been developed for continuous monitoring of dissolved oxygen (285,466). In the absence of interferences, the sensitivity of this technique ranges from 0.05 to 0.10 mg of dissolved oxygen per liter.

Oxygen-sensitive galvanic cells have been used for some time for analyses of water effluents (285). These are made of galvanic couples of an inert metal cathode (e.g., lead, zinc, or antimony) (460). The cathodic reduction of molecular oxygen results in a galvanic current proportional to the concentration of dissolved oxygen in the test solution. Changes in the pH and the conductivity of the test solution influence the oxygen measurement.

The usefulness of this type of cell is limited because the electrode system may be easily poisoned. The electrode is in direct contact with the test solution and surface-active compounds, as well as other suspended material, frequently adsorb on its surface, particularly in wastewaters. To prevent incrustation, a small amount of HCl may be added to the water by a dosing device ahead of the galvanic cell (285).

At the present time it appears that membrane electrodes offer one of the most useful techniques for analysis of dissolved oxygen in wastewaters. The unique feature of such electrode systems is that the membrane separates the electrode from the test solution and serves as a selective diffusion barrier with respect to electroactive and surface active interferences commonly found in industrial wastewaters. Two main types are presently available, the voltammetric type (189) and the galvanic cell type (190). The two types are similar in operating characteristics; however in the voltammetric type an appropriate emf source is needed, while the galvanic type is basically an oxygen energized cell.

With respect to circuitry, galvanic sensors are as simple to work with as thermocouples. All that is needed with the galvanic cell oxygen analyzer is a low impedance galvanometer or microammeter. For recording, an inexpensive galvanometer-type recorder is adequate in most cases. Also, the cell circuit may be closed with a known resistor and the potential drop fed to a pottentiometric recorder. The load resistance in the circuit has little effect on the cell sensitivity since it is in series with the large polarization resistance of the sensing electrode. At high values of external load, however, the galvanic current is affected.

A detailed discussion of the principle, operating characteristics, applicability, and limitations in the use of oxygen membrane electrodes may be found elsewhere (286,287). Similar to the membrane electrode systems discussed previously, the oxygen membrane is sensitive to the activity of the electroactive species and not necessarily the concentration. The steady-state current, $i_\infty$, is given as

$$i_\infty = [zFAD_mK_m(1/b)]a \qquad (38)$$

where $A$ is the cathodic surface area, $D_m$ is the diffusivity coefficient for oxygen in the membrane, $K_m$ is a partition coefficient for molecular oxygen at the membrane–solution interface, and $b$ is the thickness of the membrane. Concentration is related to activity by the activity coefficient,

$$a = \gamma C \qquad (39)$$

Hence,

$$i_\infty = [zFAD_mK_m(1/b)]C = \phi C \qquad (40)$$

where $\phi$ signifies the sensitivity coefficient in microamperes per unit dissolved oxygen concentration.

In the application of oxygen membrane electrodes to industrial wastewaters it is important to account for interferences caused by salting-in ($\gamma < 1.0$) and salting-out ($\gamma > 1.0$) agents, as well as for interferences from reactive species which can permeate the membrane, such as gaseous $H_2S$ and $CO_2$ (285). Industrial effluents from distilleries and pharmaceutical processes may contain salting-in agents. Wastewaters with high ionic strength, such as brine wastes, may cause a salting-out effect, which may be expressed as follows:

$$\ln \gamma = K_s I_s \qquad (41)$$

where $I_s$ is the ionic strength and $K_s$ is the salting-out coefficient. The salting-out effect on the sensitivity of the electrode system (285) is then

$$i_\infty = \phi(e^{K_s I_s})C \qquad (42)$$

The ionic strength of the test solution can be estimated by electric conductance measurements using appropriate calibration procedures. Automatic compensation for changes in ionic strength and temperature of the test solution using a simplified analog system has been described (285).

Polymeric membranes used with oxygen membrane electrodes show selective permeability to various gases and vapors. Gases reduced at the potential of the sensing electrode (e.g., $SO_2$ and halogens) cause erroneous readings, but these gases rarely exist in a free state in aqueous systems. Other gases capable of permeating plastic membranes may contaminate the sensing electrode or react with the supporting electrolyte, e.g., $CO_2$ and $H_2S$.

Another problem can arise if improper mounting of the membrane leads to trapping small air bubbles under the membrane. These bubbles cause slow electrode response and a high residual current.

## IX. CONCLUDING REMARKS

The preceding discussion provides a general survey of techniques and procedures for sampling, handling, storage, and analysis of industrial wastewaters.

Certain wastes and certain analytical methods have been examined in more detail than others. This has been determined partially by the relative significance of particular types of wastes, partially by the complexity of certain analytical problems, partially by the extent of use and availability of information on various analytical procedures, and doubtless in some small part also by the relative interests of the authors. Nonetheless, an attempt has been made to at least identify all major problems and available solutions for analysis of industrial wastes, if not discuss them all in depth.

An extensive list of sources of more detailed information and discussion of specific aspects of industrial wastewater analysis is provided by the reference list for this chapter. It is not possible within the confines of a single chapter to consider all aspects of each particular industrial wastewater situation. The reader is therefore encouraged to make use of the references cited herein to benefit fully from more complete consideration of specific analytical problems, methods, and procedures.

## APPENDIX I

### MISCELLANEOUS ORGANIC MATTER

| Substance | Occurrence | Mode of measurement | Method of analysis—Remarks | Refs. |
|---|---|---|---|---|
| Acetone | Chemical wastewater | Colorimetric | Colorimetric analysis of acetone or acetone–salicylic aldehyde or acetone furfural condensates. | 272 |
| Acrolein | Chemical wastewater | Colorimetric | The filtered sample is treated with ammonia, mixed with HCl—alcohol mixture, then treated with tryptophone in HCl. | 513 |

*(continued)*

| Substance | Occurrence | Mode of measurement | Method of analysis—Remarks | Refs. |
|---|---|---|---|---|
| Acrolein | Wastewaters | Colorimetric | Water filtered, treated with ammonia to remove iron salts, mixed with HCl and alcohol solution; interaction with 0.2 % of tryptophane causes color development. | 513 |
| Acrylonitrile | Chemical wastewater | Polarographic | Distillation with $H_2SO_4$ and $CH_3OH$ to give an axeotropic mixture boiling at 61.4°C containing 38.7 liters acrylonitrile, which is determined polarographically (sensitivity 0.1 ppm). | 134, 108 |
| Amines | Steam condensate | Colorimetric | Test for dioctadecylamine and octadecylamines; both react with methyl orange at pH 3.4–3.6 to form yellow-colored complex soluble in ethyl dichloride. | 20 |
| | Chemical wastewater | Colorimetric | Solution is treated with hypochlorite to form chloramine derivatives which are determined colorimetrically by reacting with starch–potassium iodide reagent. | 107 |
| Aniline | Chemical wastewater | Colorimetric | Color formation on reaction with chloramine-T. | 250 |
| Benzene | Chemical wastewater | Colorimetric | Benzene is converted to 1,3,-dinitrobenzene, which when reacted with methyl ethyl ketone under alkaline conditions will form a violet-colored compound (sensitivity 0.005 mg/liter and accuracy is 2–5 %). | 115 |
| Benzene | Chemical wastewater | Polarographic | Desorption by nitrogen as a carrier gas, followed by nitration and determination of the resulting $m$-dinitrobenzene polarographically. | 9 |
| Benzene hexachloride | Sewage and chemical wastewater | Spectrometric | Extraction on activated charcoal, determination as $m$-dinitrobenzene after dechlorination and nitrification by the method of Sehechtes and Hornstein, then measured by absorption spectrometry. | 182 |
| Chlortetracycline | Chemical wastewater | Fluorescence spectrometry | Compare fluorescence with standards at pH 8.5–9.0. | 397 |
| Creatinine | Infiltration water | Colorimetric | Color produced by reaction with picric acid and NaOH in the presence of sodium hexametaphosphate. | 1 |
| Dichloroethane | Chemical wastewater | Titrimetric | Sample neutralized, evaporated, burnt in a stream of air; products absorbed in NaOH, acidified, titrated with mercurous nitrate solution. | 429 |

*(continued)*

| Substance | Occurrence | Mode of measurement | Method of analysis—Remarks | Refs. |
|---|---|---|---|---|
| 2,4,-Dichloro phenoxy acetic acid | Surface waters | Colorimetric | 2,4,D heated with chromotropic acid to develop wine-red color (sensitivity 7 g/liter—range 0–30 g/liter). | 13 |
| | | Spectrometric (uv) | Direct measurement of 2,4,D uv spectra (sensitivity 30 g/liter). | 13 |
| Formaldehyde | Chemical wastewater and surface wastewater | Spectrometric | Sample is treated with phenylhydrazine hydrochloride, potassium ferricyanide, NaOH, and isopropyl alcohol for extraction. The absorption spectra in alcohol layer are determined. Acetaldehyde and ferrous iron interfere with test. | 433 |
| Formaldehyde | Plastic, tannery, and pharmaceutical wastewater | Colorimetric | Blue to greenish-blue color developed with 0.5 % of carbazole solution in conc. $H_2SO_4$ (spot test or ring test). | 160 |
| | | | Heating with chromtropic acid in 72 % $H_2SO_4$ at 60°C, develop violet color, compare with standards. | 376 |
| Furfuraldehyde | Chemical wastewater | Colorimetric | Reaction with aniline acetate and with acetophenone; another method based on benzidine reaction. | 273 |
| Heating oil and motor oil | Surface and wastewaters | Spectrometric | Liquid–liquid extraction followed by ir analysis. | 251 |
| Hexamethylene-tetramine | Chemical wastewater | Colorimetric | Hexamethylene-tetramine is hydrolyzed with dilute $H_2SO_4$ forming formaldehyde, which is measured colorimetrically using the Schiff reagent. Formaldehyde originally present can be removed by oxidation with $H_2O_2$ in alkaline solution. | 15 |
| Hydrazine | Wastewaters | Colorimetric | Sodium azide used to remove nitrite ions; 225 one part of sample treated with iodine and sodium sulfite to remove hydrazine, both parts of sample treated with $p$-dimethylaminobenzaldehyde in HCl for color development measurements based on differential absorption spectrometry of both parts of sample. | 225 |
| Hydroquinone | Wastewaters | Colorimetric | Based on reaction with 1,10-phenanthroline, sample pretreatment to eliminate interferences. | 274 |
| Lactose | Dairy wastewater | Colorimetric | After removal of proteins with sulfuric acid and sodium tungstate, alkaline copper tartarate and sodium bisulfite are added, and the cuprous copper so formed is allowed to reduce phosphomolybdic acid to molybdenum blue, which is determined colorimetrically (range 0–5000 ppm). | 201 |

(continued)

| Substance | Occurrence | Mode of measurement | Method of analysis—Remarks | Refs. |
|---|---|---|---|---|
| Lignin and ligno-sulfonic acid | Sea water | Spectrometric (uv) | Direct measurements of uv spectra in neutral and alkaline solutions. | 263 |
| Lipids | Domestic and industrial wastewaters | Gravimetric | New extraction procedures that can separate motor oils, tristeorin, sodium oleate, and various insoluble lipids. | 266 |
| Naphthalein | Industrial wastewater | Spectrometric (uv) | Extraction with heptane and differential uv spectrometric technique. | 511 |
| Naphthalene | Chemical wastewater | Colorimetric | Extraction with chloroform and NaOH, then mixing with anhydrous aluminum chloride to develop bluish-purple color in 2–3 min (range 0.2–200 ppm and accuracy + 5.6 %). | 303 |
| | | | Reaction with acetylacetone in presence of excess of ammonium salts at pH about 5.5–6.5, develops yellow color of pyridine derivative, compared with standards. | 325 |
| | | Titrimetric | Oxidation of formaldehyde by $H_2SO_4$ in presence of NaOH to form sodium formate or formation of hexamethylenetetramine by the action of ammonium chloride and caustic soda on formaldehyde. In both cases, the unused NaOH is back-titrated with HCl. | 241 |
| Nitrobenzene | Chemical wastewater | Colorimetric | Sample made alkaline, distilled, nitrobenzene is reduced to aniline, diazotized, reacted with sodium salt of 2-naphthol-3,6-disulfonic acid, and determined colorimetrically. | 170 |
| Oils and grease, chloroform-extractable matter | Oil refinery food processing | Gravimetric | Constitute a number of heavy oils, grease, rubber, certain resins, asphaltenes, and carbines. Test is based on the gravimetric determination of the residue of the chloroform extract of the sample. | 20 |
| Phthalic anhydride | Chemical wastewater | Colorimetric | Sample treated with alcohol and NaOH solution, evaporated till dryness, residue dissolved in $H_2SO_4$ and treated with acidic resorcinol; color is developed on dilution and made alkaline. | 514 |
| Proteins | Surface waters | Colorimetric | Folin–phenol reagent is reduced, in alkaline medium, by copper salt of protein, giving a blue color, the intensity of which is proportional to protein concentration (sensitivity to 0.3 mg/liter). | 357 |

*(continued)*

| Substance | Occurrence | Mode of measurement | Method of analysis—Remarks | Refs. |
|---|---|---|---|---|
| Pyridine | Oil refinery, road drainage | Colorimetric | Distillation of pyridine, buffer to pH 6–8, react with cyanogen bromide and benzidine hydrochloride and red color produced is extracted with butyl alcohol (range 0.005–1.0 ppm). | 246 |
| Pyridine | Chemical wastewater | Colorimetric | Reaction of pyridine with barbituric acid in presence of chlorocyanogen, and photocolorimetric analysis of color produced. | 123 |
| Sugars | Citrus wastewater, sulfite cellulose wastewater, beet sugar wastewater | Titrimetric | Conversion to reducing sugars by acid hydrolysis followed by heating with alkaline potassium ferricyanide; excess ferricyanide is determined idometrically. | 493 |
| Tar | Effluent gas works | Colorimetric | Extraction by chloroform in a weakly alkaline solution. Acid washing until neutral, drying over sodium sulfate, filtering, then measuring optical density (accuracy + 10 %). | 465 |
| Tar bases pyridine, quinoline, acridine | Oil refinery, coke ovens, gas works | Gravimetric | Extraction with chloroform, treated with picric acid, weighed as picrates. | 154 |
| Trichlorobenzene | Chemical wastewater | Titrimetric colorimetric | Digestion of the chloroorganic compounds in $H_2SO_4$ and $K_2Cr_2O_7$ in air, absorption of liberated chlorine by CdI solution, and determination of equivalent volume of iodine liberated titrimetrically; in cases of small amounts of liberated iodine, determinations are done colorimetrically (sensitivity = 30 g/liter and accuracy + 8 %). Interference by other chloroorganic substances. | 306 |
| Trichloroethylene | Chemical wastewater | Colorimetric | Sample is treated with pyridine and caustic soda and boiled for 5 min to develop orange color (range 1–20 ppm). | 275 |
| Trichloroethylene | Chemical wastewater | Titrimetric | Sample is treated with ammonium peroxydisulfate and nitric acid and boiled under reflux, hydrochloric acid released is titrated with $AgNO_3$ solution and potentiometric end-point detection. | 116 |
| Turpentine | Chemical wastewater | Colorimetric | Carbon dioxide passed through sample containing ethanol and phosphomolybdic acid immersed in ice; after temperature drops colorimetric determinations are done. | 271 |
| Wool wax | Textile effluents | Colorimetric | Reaction of stearine in wool wax with acetic anhydride and sulfuric acid. | 191 |

# APPENDIX II

## SYNTHETIC DETERGENTS

| Substance | Occurrence | Mode of measurement | Method of analysis—remarks | Refs. |
|---|---|---|---|---|
| Anionic surfactants | Sewage and river waters | Colorimetric | Procedure to distinguish between sulfate and sulfonate surfactants, based on hydrolyzing the sulfate detergents with boiling in sulfuric acid. | 111 |
| Anionic surfactants | Domestic and industrial effluents | Titrimetric | The titration is carried out with a standard solution of cationic surfactants (cetyl trimethylammonium bromide), in the presence of hexane; a solution of bromophenol blue or azophloxine is used as an indicator. | 128 45 |
| Anionic surfactants | Drinking water | Colorimetric | Complexation of methylene green with anionic surfactants. | 2 |
| Anionic surfactants | Surface and waste waters | — | Review article | 311 |
| Anionic surfactants | — | Polarographic | Based on dyestuff antagonist method. | 77 |
| Anionic surfactants | Surface and waste waters | — | Literature review—methods discussed are spectrometric and titrimetric methods. | 103 |
| Anionic surfactants | Textile wastewaters | Titrimetric | Surfactant is titrated with laur-2-pyridinium chloride using acid dye as indicator. | 438 |
| Anionic surfactants | — | Complexometric titration | Quantitative separation of alkyl sulfates and sulfonates as barium salt which is titrated with EDTA solution. | 185 |
| Anionic surfactants | Surface waters | Spectrometric (uv) | Separation followed by measuring uv absorption spectra. | 486 |
| Anionic surfactant | Surface waters | Ion exchange colorimetric | Separation on Amberlite IRA-68, elution with acetone and sodium hydroxide and determination by methylene blue method. | 261 |
| Anionic surfactant | Surface waters | Solvent extraction ac polarography | Separation by solvent extraction with chloroform and determination by ac polarography (range 0.07–0.028 mg ABS per 10 ml chloroform). | 228 |
| Mixed surfactants | Textile wastewaters | Colorimetric titrimetric | Review of methods for analysis for anionic, cationic, and nonionic surfactants, applicable to spent sulfate liquor. | 352 |
| Nonionic surfactants | Sewage effluents | Colorimetric | Extraction in ether, precipitation with barium phosphomolybdate in HCl–ethanol solution, digestion to form phosphates, which are measured colorimetrically by reduction to molybdenum blue. | 190 |
| Nonionic surfactants | — | Colorimetric | Surfactant reduces the optical density of dilute solutions of dichlorofluororescein in glacial acetic acid at a certain wavelength. | 437 |
| Nonionic surfactant | — | Surface tension | Surface tension measurements were made with stalagmameter at 20°C and compared with results from standard solutions. | 262 |
| Nonionic surfactant | Papermill wastewaters | Colorimetric | Separation and reaction with cobalt thiocyanate to develop a color complex. | 101 |
| Surfactants | — | Paper chromatography | Separation and identification of anionic, cationic, and nonionic surfactants by various paper chromatographic techniques. | 122 |

# APPENDIX III

## NUTRIENTS (AMMONIA, NITRITES, NITRATES AND PHOSPHATES)

| Substance | Occurrence | Mode of measurement | Method of analysis—remarks | Refs. |
|---|---|---|---|---|
| Ammonia | Water | Colorimetric | Ammonia determined in the presence of nitrates by the oxidation of nitrites to nitrates by $H_2O_2$, and subsequent distillation of ammonia from alkaline solution. | 14 |
| | | | Sample treated with chloramine-T and then allowed to react with pyridine pyrazolone reagent containing 3-methyl-1-phenyl-5-pyrazolone and a trace of the bis-pyrazolone; a purple color develops extracted with carbon tetrachloride (sensitivity 0.025 mg/liter). | 247 363 |
| | | | Based on formation of intensively blue indophenol by the reaction of ammonia, hypochlorite, and phenol; catalyzed by a manganous salt (sensitivity = 0.01 mg/liter, accuracy ± 5 %). | 69 |
| | Sea water | Colorimetric | Based on oxidation of ammonia to monochloramine; combination with phenol, extraction of resulting quinone chlorimide using $n$-hexanol, centrifugation, and determination of color intensity in the supernatant liquor. | 335 |
| Nitrites | Water and waste | Amperometric titration | Based on titration with chloramine-T (sensitivity 1 mg/liter). | 114 |
| | Surface waters | Colorimetric autoanalyzer | Hydrazine reduction test. | 156 |
| Nitrates | Natural and waste waters | Polarographic | Diffusion current measurements at $-1.2$ $V$ vs. SCE; nitrites, phosphates, and $Fe^{3+}$ interfere. | 47 |
| | Wastewaters | Colorimetric | Based on reaction of nitrates with chromotropic acid, reagent suitable for analysis in presence of formaldehyde. | 33 |
| | Waters | Spectrometric (uv) | Determine uv absorption spectra in acid solution; chlorides cause a shift in wavelength of maximum absorption measurements done at wavelength of minimum interference. | 138 |
| | Surface waters | Colorimetric | Nitrite interference prevented by preliminary treatment with urea. | 229 |
| Phosphates | Natural and wastewaters | Coulometric titration | Based on the argentometric titration of orthophosphates with $Ag^+$ ions coulometrically generated. Potentiometric and amperometric techniques were also discussed. | 89 |
| | | Colorimetric | Vanadium molybdate is used in this test and the yellow colored complex is determined colorimetrically, a technique is described to reduce interferences. | 4 150 |
| | | | Modified procedures for extraction and reduction of the phosphomolybdic acid (sensitivity 0.3 g/liter). | 436 194 |
| | | Activation analysis | Formation of phosphotungstic acid which is extracted and activated for 1 hr at a flux of $10^{13}$ neutrons/cm²/sec. | 15 |

## REFERENCES

1. Abbott, D. C., *Analyst*, **87**, 494 (1962).
2. Abbott, D. C., *Analyst*, **88**, 240 (1963).
3. Abbott, D. C., *Proc. Soc. Water Treat. Exam.*, **13**, 153 (1964).
4. Abbott, D. C., G. E. Emsden, and J. B. Harris, *Analyst*, **88**, 814 (1963).
5. Abbott, D. C., H. Egan, E. W. Hammond, and J. Thompson, *Analyst*, **89**, 480 (1964).
6. Association of British Chemical Manufacturers and Society of Analytical Chemistry Joint Committee, *Analyst*, **83**, 50 (1958).
7. Association of British Chemical Manufacturers and Society of Analytical Chemistry Joint Committee, *Analyst*, **82**, 518 (1957).
8. Association of British Chemical Manufacturers and Society of Analytical Chemistry Joint Committee, *Analyst*, **83**, 230 (1958).
9. Adamovsky, M., *Vodni Hospodarstvi*, **16**, 102 (1966), *Chem. Abstr.*, **65**, 6908f (1966).
10. Adams, C. E., Southern Pulp Paper Mfr., **1**, 12 (1959).
11. Adeney, W. E., and B. B. Dawson, *Sci. Proc. Royal Dublin Soc.*, **18**, 199 (1926).
12. Aldridge, W. N., *Analyst*, **69**, 62 (1944).
13. Aly, O. M., and S. D. Faust, *J. Amer. Water Works Assoc.*, **55**, 639 (1963).
14. Alexandrov, G. P., and W. S. Tichonova, *Lavodsk. Lab.*, (*USSR*), **26**, 57 (1960).
15. Allen, H. E., Abstracts, 155th American Chemical Society Meeting, San Francisco, April 1968.
16. Allen, L. A., *J. Soc. Chem. Ind.*, **63**, 89 (1944).
17. Allen, H., W. R. Matson, and K. H. Mancy, *J. Water Pollution Control Federation*, in press.
18. Allred, R. C., E. A. Stezkom, and R. L. Huddleston, *J. Amer. Oil Chemists Soc.*, **41**, 13 (1964).
19. Alsterberg, G., *Biochem. J.*, **159**, 36 (1925).
20. American Society of Testing Materials, *Manual on Industrial Water and Industrial Waste Water*, 2nd ed., Philadelphia, 1964.
21. American Petroleum Institute, *Manual on Disposal of Refinery Wastes—Methods for Sampling and Analyses of Refinery Wastes*, American Petroleum Institute, New York, 1957.
22. Andelman, J. B., *Proc. 140th Annual Conference, Water Pollution Control Federation, Washington, D.C.*, **1967**, p. 6.
23. Andelman, J. B., and M. J. Suess, Abstracts, 151st Meeting, American Chemical Society, Pittsburgh, March 1966.
24. Anderson, N. R., and D. N. Hume, *Advan. in Chemistry*, **73**, 30 (1968).
25. Andrews, J. F., J. F. Thomas, and E. A. Pearson, *Water and Sewage Works*, **111**, 206 (1964).
26. Anon, *Chem. Eng. News*, **10**, 46 (1965).
27. Anon, *Ind. Eng. Chem.*, **57**, 45 (1965).
28. AISI, "Annual Statistical Report," American Iron and Steel Institute, 1949.
29. American Public Health Association, American Water Works Association, and Water Pollution Control Federation, *Standard Methods for the Examination of Water and Wastewater*, 12th ed., American Public Health Association, 1965.
30. Archer, E. E., *Analyst*, **79**, 30 (1954).

31. Argonne National Laboratory, "Management of Radioactive Wastes of Argonne National Laboratory," in *Hearings of Joint Committee on Atomic Energy,* Vol. 1, 86th Congress, Superintendent of Documents, U.S. Government Printing Office, 708 (1959).
32. Ariel, M., and U. Eisner, *J. Electroanal. Chem.,* **5,** 362 (1963).
33. Armstrong, E. A., *Anal. Chem.,* **35,** 1252 (1963).
34. Arthur, R. M., *Proc. Ind. Waste Conf. 19th, Purdue Univ.,* 628 (1964).
35. Association of American Soap and Glycerine Producers, *Anal. Chem.,* **28,** 1822 (1956).
36. American Society for Testing Materials, "Symposium on Radioactivity in Industrial Water and Industrial Waste Water," *Amer. Soc. Testing Mater. Spec. Tech. Publ.,* 235 (1959).
37. Atkins, W. R. G., *J. Conseil, Conseil Perm. Intern. Exploration Mer.,* **22,** 271 (1957).
38. Baker, R. A., *J. Amer. Water Works Assoc.,* **58,** 751 (1966).
39. Baker, R. A., *J. Water Pollution Control Federation,* **34,** 582 (1962).
40. Baker, R. A., *J. Amer. Water Works Assoc.,* **56,** 92 (1964).
41. Baker, R. A., *J. Water Pollution Control Federation,* **37,** 1164 (1965).
42. Barbi, G., and S. Sandroni, *Comit. Nazl. Energia Nucl.,* **CNI-55,** 3 (1960).
43. Barker, F. B., and B. P. Robinson, "Determination of Beta Activity in Water," *U.S. Geol. Surv. Water Supply Papers,* **1696-A,** 36 (1963).
44. Barney, J. E. II, and R. J. Bertolacini, *Anal. Chem.,* **29,** 1187 (1957).
45. Barr, T., J. Oliver, and W. V. Stubbings, *J. Soc. Chem. Ind.,* **67,** 45 (1948).
46. Bartow, H., *Chem. Eng.,* **72,** 177 (1965).
47. Batlen, J. J., *Anal. Chem.,* **36,** 939 (1964).
48. Baylis, J. R., *Ind. Eng. Chem.,* **18,** 311 (1926).
49. Baxter, S. S., and V. A. Appleyard, *J. Amer. Water Works Assoc.,* **54,** 1181 (1962).
50. Bizollon, C. A., and R. Moret, *Chim. Anal.,* **46,** 273 (1964).
51. Brainina, Kh. Z., *J. Anal. Chem. (USSR),* **19,** 753 (1964).
52. Beckman Instrument Co., Fullerton, California.
53. Beier, E., *Gas-Wasserfach,* **98,** 262 (1957).
54. Belcher, R., M. A. Leonard, and T. S. West, *Talanta,* **2,** 92 (1954).
55. Bentley, E. M., and G. F. Lee, *Environ. Sci. Technol.,* **1,** 721 (1967).
56. Berck, B., *Anal. Chem.,* **25,** 1253 (1953).
57. Bertolacini, R. J., and J. E. Barney II, *Anal. Chem.,* **29,** 281 (1957).
58. Black, H., *Sewage Ind. Wastes,* **24,** 45 (1952).
59. Black, H. H., *Ind. Eng. Chem.,* **50,** 10 (1958).
60. Black, A. P., and S. A. Hannah, *J. Amer. Water Works Assoc.,* **57,** 901 (1965).
61. Blaedel, W. J., and R. H. Laessig, "Automation of the Analytical Process through Continuous Analysis," in *Advances in Analytical Chemistry and Instrumentation,* Vol. 5, C. N. Reilley, Ed., Interscience, New York, 1966, p. 69.
62. Blanchard, R. L., G. W. Leddicotte, and D. W. Moeller, *J. Amer. Water Works Assoc.,* **51,** 61 (1959).
63. Boettner, E. A., and F. I. Grunder, *Advan. Chem. Ser.,* **73,** 238 (1968).
64. Bogan, R. H., and C. N. Sawyer, *Proc. Ind. Waste Conf., 12th,* Purdue Univ. **1957,** p. 156.
65. Boltz, D. F., and M. G. Mellon, *Anal. Chem.,* **40,** 255R (1968).
66. Bollinger, L. E., *U.S. Publ. Health Serv. Publ.* **999-AP-15,** 41 (1964).

67. Bond, A. M., and M. M. Murray, *Biochem. J.*, **53**, 642 (1953).
68. Booth, R. L., J. N. English, G. N. McDermott, *J. Amer. Water Works Assoc.*, **57**, 215 (1965).
69. Bossum, J. R., and P. A. Villarruz, *J. Amer. Water Works Assoc.*, **55**, 657 (1963).
70. Bovard, P., and A. Granby, *Chim. Anal.*, **44**, 439 (1962).
71. Bramer, H. C., M. J. Walsh, and S. C. Caruso, Abstracts, 151st Meeting, American Chemical Society, Pittsburgh, March 1966.
72. Breidenbach, A. W., *U.S. Publ. Health Serv. Publ.*, **1241**, (1964).
73. Breidenbach, A. W., C. G. Gunnerson, and F. Kawahara, "Chlorinated Hydrocarbon Pesticides in Major River Basins 1957–1965," Publication of Basic Data Program, Department of Health, Education and Welfare, April 1966.
74. British Standards Institution, *Brit. Std. Inst.*, **B. S. 2690** (1953).
75. Broda, E., and T. Schonfeld, *Acta Chim. Acad. Sci. Hung.*, **50**, 49 (1966).
76. Brooks, R. R., *Cosmochim. Acta*, **29**, 1369 (1965).
77. Buchanan, G. S., and J. C. Griffith, *J. Electroanal. Chem.*, **5**, 204 (1963).
78. Bunch, R. L., and C. W. Chambers, *J. Water Pollution Control Federation*, **39**, 181 (1967).
79. Burchfield, H. P., and D. E. Johnson, *U.S. Publ. Health Serv. Rept.* 1 *and* 2 (1965).
80. Burns, E. R., and C. Marshall, *J. Water Pollution Control Federation*, **37**, 1716 (1965).
81. Burrell, D. C., *Atomic Absorption Newsletter*, Perkin Elmer Corp., Norwalk, Conn., **4**, 309 (1955).
82. Butts, P. G., A. R. Gahler, and M. G. Mellon, *Ind. Wastes*, **22**, 1543 (1950).
83. Byrd, J. F., "Combined Treatment, *Proc. Ind. Waste Conf., Purdue Univ.*, 1962, pp. 16, 109.
84. Camp, T. R., *Water and Its Impuities*, Reinhold, New York, 1963, p. 77.
85. CBEDE, *Livre De L'Eau*, Vol. 1, Centre Belge D'Etude Et De Documentation Des Eaux, Liege, Belgium, 1954.
86. Charlot, G., *Colorimetric Determination of Elements*, Elsevier, New York, 1964.
87. Cheronis, N. D., *Micro and Semimicro Methods, Technique of Oragnic Chemistry*, Vol. 6, Interscience, New York, 1954.
88. Chow, T. J., and T. G. Thompson, *Anal. Chem.*, **27**, 18 (1955).
89. Christian, G. D., E. C. Knoblock, and W. C. Purdy, *Anal. Chem.*, **35**, 1869 (1963).
90. Christman, R. F., *Proc. Water Pollution Control Federation Annual Meeting, 40th Symposium on Water Quality Analysis, New York*, 1967, p. 8.
91. Christman, R. F., and M. Ghassemi, "Progress Report to Public Health Services on Research Grant WP-00558," The University of Washington Press, Seattle, Washington, 1966.
92. Christman, R. F., and M. Ghassemi, *J. Amer. Water Works Assoc.*, **58**, 723 (1966).
93. Citron, I., H. Tsi, R. A. Day, and A. L. Underwood, *Talanta*, **8**, 798 (1961).
94. Clarke, F. E., *Ind. Eng. Chem.*, **19**, 889 (1947).
95. Cleary, E. J., *J. Amer. Water Works Assoc.*, **54**, 1347 (1962).
96. Coomber, D. I., *Proc. Soc. Water Treat. Exam.*, **12**, 100 (1963).
97. Collins, A. G., *Appl. Spectr.*, **21**, 16 (1967).

98. Collins, A. G., *U.S. Bur. Mines, Rept. Invest.*, **6641**, 18 (1965).

99. Cox, R. A., "The Physical Properties of Sea Water," *Chemical Oceanography*, Vol. 1, in J. P. Riley and G. Skirrows, Eds., Academic Press, New York, 1965, p. 73.

100. Cox, R. A., *Deep Sea Res.*, **9**, 504 (1962).

101. Crabb, N. T., and H. E. Presinger, *J. Amer. Oil Chemists' Soc.*, **41**, 752 (1964).

102. Craig, L. C., and D. Craig, in *Technique of Organic Chemistry*, 2nd ed., Vol. 3, Part I, R. Weissberger, Ed., Interscience, New York, 1956.

103. Cropton, W. G., and A. S. Joy, *Analyst*, **88**, 516 (1963).

104. Crouch, S. R., and H. V. Malmstadt, *Anal. Chem.*, **39**, 1084 (1967).

105. Crow, W. B., and F. A. Eidsness, *J. Amer. Water Works Assoc.*, **57**, 1509 (1966).

106. Csabadtm, N., *Hidrol. Kozl.*, **44**, 371 (1964).

107. Dahlgrene, G., *Anal. Chem.*, **36**, 596 (1964).

108. Daves, G. W., and W. F. Hammer, *Anal. Chem.*, **29**, 1035 (1957).

109. Davis, R. C., *J. Water Sewage Works*, **3**, 259 (1964).

110. Davis, J. B., *Ind. Eng. Chem.*, **48**, 1444 (1956).

111. Degens, P. N., Jr., H. Van Der Zee, and J. D. Kommer, *Sewage Ind. Wastes*, **25**, 24 (1953).

112. Degremont, E., *Memento Technique De L'Eau*, Degremont, Paris, 1959.

113. Delaughtor, B., *Atomic Absorption Newsletter*, Perkin Elmer Corp., Norwalk, Conn., **4**, 273 (1965).

114. Deshmukh, G. S., and S. N. Murty, *Indian J. Chem.*, **1**, 316 (1963).

115. Devlaminck, F., *Bull. Centre Belge Etude Doc. Eaux (Liege)*, **101**, 135 (1959).

116. Deyl, Z., and M. Effenberger, *Voda*, **37**, 90 (1958).

117. Dickinson, W. E., *Anal. Chem.*, **26**, 777 (1954).

118. Dobbs, R. A. and R. T. Williams, *Anal. Chem.*, **35**, 1064 (1963).

119. Dostal, H. C., and D. R. Dilley, *Bio-Science (Eng.)*, **14**, 35 (1964).

120. Donaldson, D. E., *U.S. Geol. Surv. Water Supply Paper* **550-D**, 258 (1966).

121. Drabek, R., *Chem. Tech. (Berlin)*, **9**, 77 (1957).

122. Drewry, J., *Analyst*, **88**, 225 (1963).

123. Dyathovitskaya, F. G., *Gigiena i Sanit.*, **25**, 51 (1960).

124. Eckenfelder, W. W., and E. Barnhart, "Removal of Synthetic Detergents from Laundry and Laundromat Wastes," Research Report No. 5, New York State Water Pollution Control Board, (March 1960).

125. Eckfeldt, E. L., and E. W. Shaffer, *Anal. Chem.*, **36**, 2008 (1964).

126. Edge, R. A., R. R. Brooks, L. H. Ahrens, and S. Amdurer, *Geochim. Cosmochim. Acta*, **15**, 337 (1959).

127. Edwards, G. P., J. P. Pfafflin, L. H. Schwartz, and P. M. Lauren, *J. Water Pollution Control Federation*, **34**, 1112 (1962).

128. Edwards, G. P., and M. E. Ginn, *Sewage Ind. Wastes*, **26**, 945 (1954).

129. Effenberger, M., *Sci. Papers, Inst. Chem. Technol. (Prague), Technol. Water*, **6**, 471 (1962).

130. Egan, T., E. W. Hammond, and J. Thomson, *Analyst*, **89**, 175 (1964).

131. Eisner, U., M. Rottschaffer, F. J. Berlandi, and H. B. Mark, Jr., *Anal. Chem.*, **39**, 1466 (1967).

132. Eisenman, G., in *Advances in Analytical Chemistry and Instrumentation*, Vol. 4, C. N. Reilley, Ed., Interscience, New York, 1965, p. 213.

133. Ellis, David W., and D. R. Demers, *Advan. Chem. Ser.*, **73**, 326 (1968).

134. Ellis, M. M., B. A. Westfall, and M. D. Ellis, *U.S. Fish Wildlife Serv. Res. Rept.,* **9** (1948).
135. Elwell, W. T., and J. A. F. Gidley, *Atomic Absorption Spectrophotometry,* Pergamon Press, Oxford, 1966.
135a. Emerson, E., *J. Org. Chem.,* **8,** 417 (1943).
135b. Emerson, E., and K. Kelly, *J. Org. Chem.,* **13,** 532 (1948).
136. Erdey, L, and F. Szabadvary, *Acta. Chim. Acad. Sci., Hung.,* **4,** 325 (1954).
137. Ettinger, M. B., and C. C. Ruchhoft, *Anal. Chem.,* **20,** 1119 (1948).
138. Fadrus, H., and J. Maly, *Z. Anal. Chem.,* **202,** 164 (1964).
139. Fahrenfort, J., *Spectrochim. Acta.,* **17,** 698 (1961).
140. Fair, G. M., *Sewage Works J.,* **8,** 430 (1936).
141. Fair, G. M., *J. New Engl. Water Works Assoc.,* **47,** 248 (1933).
142. Fair, G. M., J. C. Geyer, *Water Supply and Waste Water Disposal,* New York, 1954.
143. Fairing, I. D., and F. R. Short, *Anal. Chem.,* **28,** 1827 (1956).
144. Faust, S. D., and O. M. Aly, *J. Amer. Water Works Assoc.,* **54,** 235 (1962).
145. Faust, S. D., and N. E. Hunter, *J. Amer. Water Works Assoc.,* **57,** 1028 (1965).
145a. Faust, S. D., and E. W. Mikulewicz, *Water Res.* **1,** 405 (1967).
145b. Faust, S. D , and E. W. Mikulewicz, *Water Res.,* **1,** 509 (1967).
146. Feigl, F., R. Belcher, and W. I. Setephen, in *Advances in Analytical Chemistry and Instrumentation,* Vol. 2, C. N. Reilley, Ed., Interscience, New York, 1963, p. 1.
147. Ferrari, A., *Ann. N. Y. Acad. Sci.,* **87,** 792 (1960).
148. Fishman, M. J., and M. R. Midget, *Advan. Chem. Ser.,* **73,** 230 (1968).
149. Fleszar, B., *Chem. Anal.,* **9,** 1075 (1964).
150. Fogg, D. N., and N. T. Wilkinson, *Analyst,* **83,** 406 (1958).
151. Folkendorf, E., *Kernenergie,* **7,** 108 (1964).
152. Fonselius, S. G., E. Kovoleff, *Bull. Inst. Oceanog.,* **61,** 1 (1963).
153. Foulke, D. G., *Metal Finish,* **47,** 58 (1949).
154. Fox, J. J., and J. H. Gauge, *J. Soc. Chem. Ind.,* **39,** 260T (1920).
155. Frant, M. S., and J. W. Ross, *Science,* **154,** 1553 (1966).
155a. Frant, M. S., *Plating,* **54,** 702 (1967).
155b. Frant, M. S., and J. W. Ross, *Anal. Chem.,* **40,** 1169 (1968).
156. Frasier, R. E., *J. Amer. Water Works Assoc.,* **55,** 624 (1963).
157. Friedrich, A., E. Kuhaas, and R. Schurch, *Z. Physiol. Chem.,* **2/6,** 68 (1933).
158. Fritz, J. S., and S. K. Karakker, *Anal. Chem.,* **32,** 957 (1960).
159. Fritz, J. S., and S. S. Yamamura, *Anal. Chem.,* **27,** 1461 (1955).
160. Furst, K., *Mikrochemie,* **33,** 348 (1948).
161. Gates, W. E., K. H. Mancy, F. Shafie, and F. Pohland, *Proc. Purdue Ind. Waste Conf.,* **1966,** p. 238.
162. Gardner, D. G., R. F. Muraca, and E. J. Serfass, *Plating,* **43,** 743 (1956).
163. Gebauer, H., and S. Muller, *Atomwirtschoft,* **1,** 487 (1962).
164. Gellman, I., and H. Heukelekian, *Sewage Ind. Wastes,* **23,** 1267 (1951).
165. Gerald, F. A., *The Human Senses,* Wiley, New York, 1953.
166. Gerischer, H., *Z. Physik. Chem. (Leipzig),* **202,** 302 (1953).
167. Geyer, R., and W. Syring, *Z. Chem.,* **6,** 92 (1966).
168. Gibbs, H. D., *J. Biol. Chem.,* **72,** 649 (1927).
169. Gohlke, R. S., *Anal. Chem.,* **34,** 1332 (1962).
170. Golubeva, M. T., *Lab. Delo,* **1,** 6 (1964).

171. Goodenkamf, A., and J. Erdei, *J. Amer. Water Works Assoc.,* **56,** 600 (1964).
172. Gorbach, S., and F. Ehrenberger, *Z. Anal. Chem.,* **181,** 106 (1961).
173. Gotlieb, E., and J. T. Pelczar, Jr., *Bacteria Rev.,* **15,** 55 (1951).
174. Grande, J. A., and J. Beukenkamp, *Analyt. Chem.,* **28,** 1497 (1956).
175. Gunther, F. A., T. A. Miller, and T. E. Jenkins, *Anal. Chem.,* **37,** 1186 (1965).
176. Gurnham, C. F., *Industrial Wastewater Control,* Academic Press, New York, 1965.
177. Haase, L. W., and E. V. Gessellschaft, *Standard Methods for the Examination of Water, and Sludge,* (in German), Verlag Chemie, Wienheim, 1954.
178. Habashi, G., *Mikrochim. Acta,* **23,** 233 (1960).
179. Hallbach, P. F., "Radionuclide Analysis of Environmental Samples," A Laboratory Manual of Methodology, *U.S. Publ. Health Service, Div. Radiological Health, Rept.* **R50-6,** (1959).
180. Hammerton, C., *J. Appl. Chem.,* **5,** 517 (1955).
181. Hampson, B. L., *Analyst,* **89,** 651 (1964).
182. Hancock, W., and E. Q. Law, *Analyst,* **80,** 665 (1955).
183. Haney, P. D., and J. Schmidt, *U.S. Publ. Health Serv. Tech. Rept.,* **W58-2,** 133 (1958).
184. Hanna, George P., P. J. Weaver, W. D. Sheets, and R. H. Gerhold, *Water and Sewage Works,* **111,** 478 (1964).
185. Harada, T., and W. Kimura, *J. Japan Oil Chemists' Soc.,* **7,** 77 (1958).
186. Harrick, N. J., *J. Phys. Chem.,* **64,** 1110 (1960).
187. Harrick, N. J., Ann, *N.Y., Acad. Sci.,* **101,** 928 (1963).
188. Hayashi, K., T. Danzuka, and K. Ueno, *Talanta,* **4,** 244 (1960).
189. Hazen, A., *J. Amcr. Chem. Soc.,* **12,** 427 (1892).
190. Heatley, N. G., and E. J. Page, *Water Sanit. Eng.,* **3,** 46 (1952).
191. Heidler, K., *Chem. Tech.,* **8,** 160 (1956).
192. Heinerth, E., *Tenside,* **3,** 109 (1966).
193. Heller, K., G. Kuhla, and F. Machek, *Mikrochemie,* **18,** 193 (1935).
194. Henriksen, A., *Analyst,* **88,** 898 (1963).
195. Henderson, A. D., and J. J. Baffa, *Proc. Amer. Soc. Civ. Eng.,* **80,** 494 (1954).
196. Hetman, J., *J. Appl. Chem.,* **10,** 16 (1960).
197. Heukelekian, H., *Ind. Eng. Chem.,* **42,** 647 (1950).
198. Higashiura, M., *Kagaku To Kogyo (Tokyo),* **38,** 306 (1964).
199. Higashiura, M., *Kagaku To Kogyo (Tokyo),* **40,** 90 (1966).
200. Hill, R. H., and L. K. Herdon, *Sewage Ind. Wastes,* **24,** 1389 (1952).
201. Hindin, E., *Sewage Ind. Wastes,* **25,** 188 (1953).
202. Hinkle, M. E., and W. A. Koehler, "Investigations of Coal Mine Drainage," West Virginia University Engineering Experiment Station, 1944, 1946.
203. Hissel, J., and J. Price, *Bull. Centre Belge Etude Doc. Eaux (Liege),* **44,** 76 (1959).
204. Hiser, L L., and A. W. Busch, *J. Water Pollution Control Federation,* **36,** 505 (1964).
205. Hoak, R. D., H. C. Bramer, and S. C. Caruso, "Recovery of Organics by Continuous Countercurrent Extraction," Mellon Institute, Pittsburgh, 1962.
206. Horton, R. K., M. Pachelo, and M. F. Santana, *Proc. Inter-Amer. Reg. Conf., Sanitary Eng., Caracus, Venezuela,* **1946,** p. 205.
207. Horton, J. P., J. D. Molley, and H. C. Bays, *Sewage Ind. Wastes,* **28,** 70 (1956).

208. Houlihau, J. E., and P. E. L. Farina, *Sewage Ind. Wastes,* **24,** 157 (1952).
209. Hrivnak, J., *Vodni Hospodarstvi,* **14,** 394 (1964); *Chem. Abstr.,* **62,** 8822h (1965).
210. Hurford, T. R., and D. F. Boltz, *Abstracts of Papers, 154th Meeting, American Chemical Society,* Chicago, 1967.
211. Iguchi, A., *Bull. Chem. Soc. Japan,* **31,** 600 (1958).
212. Institution of Water Engineers, "Approved Methods for the Physical and Chemical Examination of Water," 3rd ed., W. Heffer & Sons, Ltd., Cambridge, England, 1960.
213. Ingols, R., and P. E. Murray, *Water Sewage Works,* **95,** 113 (1948).
214. Inczedy, J., *Analytical Applications of Ion Exchangers,* Pergamon Press, 1966, London, p. 151.
215. Inczedy, J., *Wasserchemie,* **9,** 80 (1964).
216. Isenhour, T. L., C. A. Evans, and G. H. Morrison, *Proc. Intern. Conf. Modern Trends in Activation Analysis, Texas A & M University, College Station,* **123,** 236 (1965).
217. Irudayasamy, A., and A. R. Natarajan, *Analyst,* **90,** 503 (1965).
218. Jaeger, K., and W. Niemitz, *Sewage Ind. Wastes,* **25,** 631 (1953).
219. Jenkins, D., and L. L. Medsker, *Anal. Chem.,* **36,** 610 (1964).
220. Joensson, G., *U.S. At. Energy Comm.,* **AE-105,** 9 (1963).
221. Johnson, D. W., and H. O. Halvorson, *J. Bacteriol.,* **42,** 145 (1941).
222. Jones, H., *The Science of Color,* Thomas Y. Crowell Co., New York, 1952.
223. Joyner, T., J. S. Finley, *Atomic Absorption Newsletter,* Perkin Elmer Corp., Norwalk, Conn., **5,** 21 (1966).
224. Kalalova, E., *Sb. Vysoke Skoly Chem. Technol. Praze, Fak. Technol. Vody,* **5,** 55 (1962).
225. Kalinia, N. M., *Emergetik,* **31,** 802 (1965); *Chem. Abstr.,* **62,** 7503c (1965).
226. Kagi, J. H. R., and B. L. Vallee, *Anal. Chem.,* **30,** 1951 (1958).
227. Kahn, Lloyd, and Cooper H. Wayman, *Anal. Chem.,* **36,** 1340 (1964).
228. Kahn, H. L., *Advan. Chem. Ser.,* **73,** 183 (1968).
228a. Kambara, T., and K. Hasebe, *Bunseki Kagaku,* **14,** 491 (1965).
229. Kamphake, L. J., S. A. Hannah, and J. M. Cohen, *Water Res.,* **1,** 205 (1967).
230. Katlofsky, B., and R. E. Keller, *Anal. Chem.,* **35,** 1665 (1963).
231. Kelley, T. F., *Anal. Chem.,* **37,** 1078 (1965).
232. Kemula, W., Z. Kublick, and Z. Galus, *Nature,* **184,** 1795 (1959).
233. Key, A., *Gas Works Effluents and Ammonia,* 2nd Ed., Institute of Gas Engineers, London, 1956.
234. Kirkbright, G. F., T. S. West, and C. Woodward, *Anal. Chem.,* **37,** 137 (1965).
235. Kirkbright, G. F., T. S. West, and C. Woodward, *Anal. Chem.,* **38,** 1558 (1966).
236. Kleeman, C. R., E. Taborsky, and F. H. Epstein, *Proc. Soc. Exptl. Biol.,* **91,** 480 (1956).
237. Klein, L. J., *Proc. Inst. Sewage Purif.,* 174 (1941).
238. Klein, L. J., *J. Inst. Sewage Purif.,* **2,** 153 (1950).
239. Kabayashi, S., and G. F. Lee, *Anal. Chem.,* **36,** 2197 (1964).
240. Koch, R. C., *Activation Analysis Handbook,* Academic Press, New York, 1960.
241. Kolthoff, I. M., and V. A. Stenger, *Volumetric Analysis,* Vol. 2, 2nd ed., Interscience, New York, 1942.
241a. Kolthoff, I. M., W. D. Harris, and G. Matsuyama, *J. Amer. Chem. Soc.,* **66,** 1782 (1944).

242. Kopp, J. F., and R. C. Kroner, *Appl. Spectr.*, **19**, 155 (1965).

243. Kjeldahl, J. *Z. Anal. Chem.*, **22**, 366 (1883).

244. Kullgren, C., *Svensk Kem. Tidskr.*, **43**, 99 (1931).

245. Kreg, J., and K. H. Szekielda, *Z. Anal. Chem.*, **207**, 388 (1965).

246. Kroner, R. C., M. B. Ettinger, and W. A. Moore, *Anal. Chem.*, **24**, 1877 (1952).

247. Kruse, J. M., and M. G. Mellon, *Anal. Chem.*, **25**, 1188 (1953).

248. Kruse, J. M., and M. G. Mellon, *Sewage Ind. Wastes*, **23**, 1402 (1951).

249. Kruse, J. M., and M. G. Mellon, *Sewage Ind. Wastes*, **24**, 1254 (1952).

250. Kuper, A. I., *Gigiena i Sanit.*, **22**, 61 (1957).

251. Ladendorf, P., *Vom Wasser*, **29**, 119 (1962).

252. Lakanen, E., *Atomic Absorption Newsletter*, Perkin Elmer Corp., Norwalk, Conn., **5**, 17 (1966).

253. Laitinen, H. A., *Natl. Bur. Std. (U.S.) Monograph*, **100**, 75 (1967).

254. Lamar, W. L., and P. G. Drake, *J. Amer. Water Works Assoc.*, **47**, 563 (1955).

255. Lamer, W. L., D. F. Goerlitz, and L. M. Law, *Advan. Chem. Ser.*, **60**, 305 (1966).

256. Leconite, M., *Rev. Gen. Thermique*, **4**, 629 (1965).

257. Leddicotte, G. W., *U.S. At. Energy Comm.*, **65-372-2**, 17 (1966).

258. Lederer, V. V., *Chem. Listy*, **41**, 230 (1947).

259. Lee, E. W., and W. J. Oswald, *Sewage Ind. Wastes*, **26**, 9, 1097 (1954).

260. Leibnitz, E., U. Behrens, and A. Gabert, *Wasserwirtsch.-Wassertech.*, **9**, 69 (1958).

261. LePeintre, C., and C. Romens, *Compt. Rend.*, **261**, 452 (1965).

262. Levchenko, O. N., A. D. Khudyakova, and N. D. Gadrivola, *Lavodsk. Lab.*, **27**, 408 (1961).

263. Lindberg, A., *Vattenhygien*, **19**, 106 (1963).

264. Lindenbaum, A., J. Schubert, and W. D. Armstrong, *Anal. Chem.*, **20**, 1120 (1948).

265. Lingane, J. J., *Anal. Chem.*, **39**, 881 (1967).

266. Loehr, R. C., and G. C. Higgins, *Air Water Pollution*, **9**, 55 (1965).

267. Longwell, J., and W. D. Maniece, *Analyst*, **80**, 167 (1955).

268. Loveridge, B. A., G. W. E. Milner, G. A. Barnett, A. Thomas, and W. M. Henry, *A.E.R.E.* (Harwell), **R-3323** (1960).

269. Ludzack, F. J., and M. B. Ettinger, *Proc. Ind. Wastes Conf., 14th, Purdue Univ.*, **1959**, p. 14.

270. Luré, Yu Yu, and Z. V. Nikolaeva, *Zavodsk. Lab.*, **30**, 937 (1964); *Chem. Abstr.*, **61**, 13035e (1964).

271. Luré, Yu Yu, and V. A. Panova, *Zavodsk. Lab.*, **29**, 293 (1963).

272. Luré, Yu Yu, and Z. W. Nikolajeva, *Zavodsk. Lab.*, **21**, 410 (1955).

273. Luré, Yu Yu, and V. A. Panova, *Zavodsk. Lab.*, **28**, 281 (1962).

274. Luré, Yu Yu, and Z. V. Nikolaeva, *Zavodsk. Lab.*, **31**, 802 (1965); *Chem. Abstr.*, **63**, 12859a (1965).

275. Lyne, F. A., and T. McLachlan, *Analyst*, **74**, 513 (1949).

276. Ma, T. S., R. E. Lang, and J. D. McKinley, *Mikrochim. Acta.*, **36**, 57, 368 (1957).

277. Macchi, G., *J. Electroanal. Chem.*, **9**, 920 (1965).

278. MacKichan, K. A., *J. Amer. Water Works Assoc.*, **53**, 1211 (1961).

279. Maienthal, E. J., and J. K. Jaylor, *Advan. Chem. Ser.*, **73**, 172 (1968).

280. Malissa, H., and E. Schoffmann, *Microchim. Acta*, **40**, 187 (1955).

281. Malo, B. A., and R. A. Baker, *Advan. Chem. Ser.*, **73**, 149 (1968).
282. Malz, W., *Foederation Europaeischer Gewaesserschutz Informationsbl.*, **11**, 19 (1964); *Chem. Abstr.*, **64**, 1808g (1966).
283. Mancy, K. H., Unpublished work, School of Public Health, University of Michigan, 1968.
284. Mancy, K. H., and T. Jaffe, *U.S. Public Health Serv. Publ.*, **999-WP-37**, 1966.
285. Mancy, K. H., and D. A. Okun, *Anal. Chem.*, **32**, 108 (1960).
286. Mancy, K. H., D. A. Okun, and C. N. Reilley, *J. Electroanal. Chem.*, **4**, 65 (1962).
287. Mancy, K. H., and W. C. Westgarth, *J. Water Pollution Control Federation*, **34**, 1037 (1962).
288. Manecke, G., C. Bahr, and C. Reich, *Angew. Chem.*, **73**, 299 (1961).
289. Mansell, R. E., and H. W. Emmel, *Atomic Absorption Newsletter*, Perkin Elmer Corp., Norwalk, Conn., **4**, 365 (1965).
290. Marcie, F. J., *Environ. Sci. Technol.*, **1**, 164 (1967).
291. Mark, H. B., private communications, Chemistry Department, The University of Michigan, 1967.
292. Marten, J. F., E. W. Cantanzaro, and D. R. Grady, *Proc. Technicon Symp. Automation Anal. Chem., New York City*, p. 259, October 1966.
293. Masselli, J. W., N. W. Masselli, and M. G. Burford, *New England Interstate Water Pollution Control Commission*, 1958.
294. Masselli, J. W., N. W. Masselli, and M. G. Burford, *New England Interstate Water Pollution Control Commission*, 1959.
295. Matson, W. R., D. K. Roe, and D. E. Carritt, *Anal. Chem.*, **37**, 1594 (1965).
296. May, D. S., Jr., and G. H. Dunstan, *Proc. Purdue Ind. Waste Conf.*, **15**, 321 (1963).
297. McCoy, J. W., *Anal. Chim. Acta*, **6**, 259 (1956).
298. McKee, J. E., and H. W. Wolf, *Water Quality Criteria*, Publication No. 3-A, State of California, State Water Quality Board, Sacramento, 1954.
299. McKeown, J. J., "Procedures for Conducting Mill Effluent Surveys," Technical Bulletin 183, National Council for Steam Improvement, 1965.
300. McKinney, D. S., and A. M. Amorosi, *Ind. Eng. Chem.*, **16**, 315 (1944).
301. McKinney, E., and J. M. Symons, *Sewage Ind. Wastes*, **31**, 549 (1959).
302. McNutt, N. W., and R. H. Maier, *Anal. Chem.*, **34**, 276 (1962).
303. Medin, A. L., and L. K. Herndon, *Sewage Ind. Wastes*, **24**, 1260 (1952).
304. Meinke, W. W., *Science*, **121**, 177 (1955).
305. Meites, L., *Handbook of Analytical Chemistry*, McGraw-Hill, New York, 1963.
306. Meleshchenko, K. F., *Gigiena i Sanit.*, **25**, 54 (1960).
307. Mellon, M. G., *Analytical Absorption Spectroscopy*, Wiley, New York, 1950, Chapter 9.
308. Melpolder, F. W., C. W. Warfield, and C. E. Headington, *Anal. Chem.*, **25**, 1453 (1953).
309. Mentink, A. F., "Instrumentation for Water Quality Determination," *ASCE Water Resources Eng. Conf., Reprint* **153** (1965).
310. Meyer, H. J., and C. Brack, *Chem. Eng. Techol.*, **22**, 545 (1950).
311. Michelsen, E., and E. Marki, *Mitt Gebiete Lebensm. Hyg.*, **52**, 557 (1961).
312. Mills, E. J., Jr., and V. T. Stack, Jr., *Proc. Ind. Waste Conf., 9th, Purdue Univ.*, **57**, 1954.

313. Mills, E. J., and V. T. Stack, Jr., *Proc. Ind. Waste Conf., 8th, Purdue Univ.,* **273,** 1953.
314. Milton, R. F., *Analyst,* **74,** 54 (1949).
315. Mitchell, L. E., *Advan. Chem. Ser.,* **60,** 1 (1966).
316. Mohlman, F. W., and F. P. Edwards, *Ind. Eng. Chem,* **3,** 119 (1963).
317. Moggio, W. A., *Proc. Amer. Soc. Civil Engs.,* **80,** 420 (1954).
318. Molof, A. H., and N. S. Zaleiko, *Proc. Purdue Ind. Waste Conf., 19th, Purdue Univ.,* **1964,** p. 56.
319. Montgomery, H. A. C., and N. S. Thom, *Analyst,* **87,** 689 (1962).
320. Moore, E. Q., H. A. Thomas, Jr., and W. B. Snow, *Sewage Ind. Wastes,* **22,** 1343 (1950).
321. Moore, W. A., R. C. Kroner, and C. C. Ruchhoft, *Anal. Chem.,* **21,** 953 (1949).
322. Moore, W. A., F. J. Ludzack, and C. C. Ruchhoft, *Anal. Chem.,* **23,** 1297 (1951).
323. Morris, R. L., *Anal Chem.,* **24,** 1376 (1952).
324. Myers, L. S., and A. H. Brush, *Anal. Chem.,* **34,** 342 (1962).
325. Nash, T., *Biochem. J.,* **55,** 416 (1953).
326. National Bureau of Standards, "Recommendations of the National Committee on Radiation Protection and Measurements," Handbook 49, *U.S Government Printing House,* November 1951.
327. National Bureau of Standards, "Radiological Monitoring Methods and Instruments," Handbook 51, U.S. Government Printing House, April 1952.
328. National Bureau of Standards, "Maximum Permissible Amounts of Radioisotopes in the Human Body and Maximum Permissible Concentrations in Air and Water," Handbook 52, U.S. Government Printing House, March 1953.
329. National Bureau of Standards, "Permissible Dose from External Sources of Ionizing Radiation," Handbook 59, U.S. Government Printing House, September 1954.
330. Naucke, W., and F. Tarkmann, *Brennstoff-Chem.,* **45,** 263 (1964); *Chem. Abstr.,* **62,** 322b (1965).
331. Navone, R., and W. D. Fenninger, *J. Amer. Water Works Assoc.,* **59,** 757 (1967).
332. Nemerow, N. L., *Theories and Practices of Industrial Waste Treatment,* Addison-Wesley, Reading, Mass., 1963.
333. Nemtsova, L. I., A. D. Semenov, and V. G. Datsko, *Gidrokhim. Materialy,* **38,** 150 (1964); *Chem. Abstr.,* **62,** 15902x (1965).
334. Newell, J. C., R. J. Mazaika, and W. J. Cook, *J. Agr. Food Chem.,* **6,** 669 (1958).
335. Newell, B., and G. D. Pont, *Nature,* **201,** 36 (1964).
336. Nusbaum, I., *Sewage Ind. Wastes,* **25,** 311 (1953).
337. Oak Ridge National Laboratory and Robert A. Taft Sanitary Engineering Center, PHS, "Report of the Joint Program of Studies on the Decontamination of Radioactive Waters," *U.S. At. Energy Comm., Rept.,* **ORNL-2557,** 45 (1959).
338. O'Brien, J. E., and J. Fiore, *Wastes Eng.,* **33,** 352 (1962).
339. O'Brien, J. E., and J. Fiore, *Wastes Eng.,* **33,** 128 (1962).
340. O'Kelley, G. D., "Application of Computers to Nuclear and Radio Chemistry," *U.S. At. Energy Comm. Rept.,* **NASNS 3017,** p. 156 (1962).

341. Otsuki, A., and T. Hanya, *Nippon Kagaku Zasshi,* 84, 798 (1963); *Chem. Abstr.,* 60, 2640e (1964).
342. Orford, H. E., and W. T. Ingram, *Sewage Ind. Wastes,* 25, 419 (1953).
343. Orford, H. E., and W. T. Ingram, *Sewage Ind. Wastes,* 25, 424 (1953).
344. Oppenheimer, C. H., E. F. Corcoran, and J. Van Arman, *Limnol. Oceanog.,* 8, 478 (1963).
345. Palange, R. C., and S. A. Megregian, *J. Sanit. Eng., Amer. Soc. Civil Engs.,* 84, 1606 (1958).
346. Palin, A. T., *Water Water Eng.,* 59, 341 (1955).
347. Parsons, M. L., W. J. McCarthy, and J. D. Winefordner, *J. Chem. Ed.,* 44, 214 (1967).
348. Pearson, E., and C. N. Sawyer, *Proc. Ind. Waste Conf., 5th, Purdue Univ.,* 1949, p. 26.
349. Pichler, H., and H. Schulz, *Brennstaff-Chem.,* 39, 148 (1958); *Chem. Abstr.,* 57, 1978 (1959).
350. Pickhardt, W. P., A. M. Oemler, and I. Mitchell, Jr., *Anal. Chem.,* 27, 1784 (1955).
351. Pillion, E., *J. Gas Chromatog,* 3, 238 (1965).
352. Pitter, P., *Sci. Papers Inst. Chem. Technol., Water (ČSSR),* 6, 547 (1962).
353. Ponomarenko, A. A., and L. N. Amelina, *Zh. Analit. Khim.,* 18, 1244 (1963); *Chem. Abstr.,* 60, 3494h (1954).
354. Popel, F., and C. Wagner, "Determination of Organic Carbon," unpublished manuscript, *Forschungs und Entwicklungsinstutut fur Industrie—und Siedlungs-wasserwirtschaft sowie Abfallwertschaft e.B.,* Stuttgart, Germany.
355. Porges, R., and E. J. Struzeski, *J. Water Pollution Control Federation,* 33, 167 (1961).
356. Potter, E. C., and J. F. Moresky, "The Use of Cation-Exchange Resin in the Determination of Copper and Iron Dissolved at Very Low Concentrations in Water," in *Ion Exchange and Its Applications,* Society of Chemical Industry, London, 1955, p. 93.
357. Povoledo, D., and M. Gerletti, *Mem. Inst. Ital. Idrobiol. Dott. Marco de Marchi,* 15, 153 (1962).
358. Powell, S. T., *Ind. Eng. Chem.,* 46, 112A (1954).
359 Pohl, F. A., *Z. Anal. Chem.,* 197, 193 (1963).
360. Pomeroy, R., *Sewage Works J.,* 8, 572 (1936).
361. Potter, E. C., and G. E. Everitt, *J. Appl. Chem.,* 9, 642 (1959).
362. Precision Scientific Co., Chicago. Illinois.
363. Prochazkova, L., *Anal. Chem.,* 36, 865 (1964).
364. Promeroy, R., *Petroleum Eng.,* 15, 156 (1944).
365. Prussin, S. G., T. A. Harris, and T. M. Hollander, *Proc. Intern. Conf. Mod. Trends Activation Anal., Texas A & M University, College Station,* 123, 132 (1965).
366. Pungor, E, K. Toth, and J. Havas, *Acta Chim. Acad. Sci. Hung.,* 41, 239 (1964).
367. Pungor, E., and J. Havas, *Acta Chim. Acad. Sci. Hung.,* 50, 77 (1966).
368. Pungor, E., K. Toth, and J. Havas, *Acta Chim. Acad. Sci. Hung.,* 48, 17 (1966).
369. U.S. Department of Public Works, *Public Works,* 85 (1954).

370. Rainwater, F. H., and L. L. Thatcher, *U.S. Geol. Surv., Water, Supply Paper,* **1454** (1960).
371. Rambow, C. A., *J. Amer. Water Works Assoc.,* **55,** 1037 (1963).
372. Rayonier's, Inc., *Paper Mill News,* **82,** 12 (1959).
373. Rechnitz, G. A., M. R. Kiesz, and S. B. Zamochnick, *Anal. Chem.,* **38,** 973 (1966).
374. Rechnitz, G. A., M. R. Kresz, *Anal. Chem.,* **38,** 1786 (1966).
375. Reilley, C. N., and D. T. Sawyer, *Experiments for Instrumental Methods,* McGraw-Hill, New York, 1961.
376. Reynolds, J. G., and M. Irwin, *Chem. Ind. Rev.,* **419** (1948).
377. Richter, H. G., and A. S. Gillespie, Jr., *Anal. Chem.,* **34,** 1116 (1962).
378. Rideal, S., and G. Stewart, *Analyst,* **26,** 141 (1901).
379. Rider, B. F., and M. F. Mellon, *Ind. Eng. Chem. (Anal.),* **18,** 96 (1946).
380. Riehl, M. L, and E. G. Will, *Sewage Ind. Wastes,* **22,** 190 (1950).
381. Riehl, M. L., *Sewage Works J.,* **20,** 629 (1948).
382. Riley, J. P., and P. Shinhaseni, *Analyst,* **83,** 299 (1958).
383. Robinson, J. W., "Atomic Absorption Spectroscopy," Marcell Dekker, Inc., New York, 1966.
384. Robertson, G., *Analyst,* **89,** 368 (1964)
385. Roderick, W. R., *J. Chem. Ed.,* **43,** 510 (1966).
386. Rogers, C. J., C. W. Chambers, and N. A. Clarke, *Anal. Chem.,* **38,** 1853 (1967).
387. Rosen, A. A., R. T. Skeel, and M. B. Ettinger, *J. Water Pollution Control Federation,* **35,** 777 (1963).
388. Rosen, A. A., J. B. Peter, and F. M. Middleton, *J. Water Pollution Control Federation,* **34,** 7 (1962).
389. Rosen, A. A., and F. M. Middleton, *Anal. Chem.,* **31,** 1729 (1959).
390. Rosen, A. A., personal communication, Cincinnati Water Pollution Control Laboratory, FWPCA, USDI, 1966.
391. Rosen, A. A., and F. M. Middleton, *Anal. Chem.,* **27,** 790 (1955).
392. Ross, J. W., *Science,* **156,** 1378 (1967).
393. Rossum, J. R, *Anal. Chem.,* **21,** 631 (1949).
394. Rudolfs, W., *Industrial Waste Treatment,* Reinhold, New York, 1953.
395. Rudolfs, W., and W. D. Hanlon, *Sewage Ind. Wastes,* **23,** 1125 (1951).
396. Rush, R. M., and J. H. Yoe, *Anal. Chem.,* **26,** 1345 (1954).
397. Shabunin, I. I., and V. I. Lavrenchuk, *Gigiena i Sanit.,* **30,** 63 (1965); *Chem. Abstr.,* **62,** 14320f (1965).
398. Sanderson, W. W, and G. B. Ceresia, *J. Water Pollution Control Federation,* **37,** 1177 (1965).
399. Sawicki, E., T. W. Stanley, J. Pfaff, and A. D. Amico, *Talanta,* **10,** 641 (1963).
400. Sax, N. I., M. Beigel, and J. C. Daly, *Ann. Rept. N.Y. Dept. Health Res. Lab.,* **1962,** 75.
401. Schatze, T. C., *Sewage Works J.,* **17,** 497 (1945).
402. Schechter, M. S., S. B. Soloway, R. A. Hayes, and H. L. Haller, *Ind. Eng. Chem.,* **17,** 704 (1945).
403. Schechter, M. S., and I. Hornstein, *Anal. Chem.,* **24,** 544 (1952).
404. Schilt, A. A., *Anal. Chem.,* **30,** 1409 (1958)
405. Scholz, L., *Vom Wasser,* **30,** 143 (1963).

406. Schoniger, W., in *Advances in Analytical Chemistry and Instrumentation*, Vol. I, 199, C. N. Reilley, Ed., Interscience, New York, (1965).
407. Schmidt, H., and M. von Stackelbert, *Modern Polarographic Methods*, Academic Press, New York, 1963.
408. Schulz, N. F., *Anal. Chem.*, **25**, 1762 (1953).
409. Schweitzer, G. K., and W. Van Willis, *Advances in Analytical Chemistry and Instrumentation*, Vol. 5, C. N. Reilley, Ed., Interscience, New York, 1966, p. 169.
410. Serfass, E. J., R. B. Freeman, B. F. Dodge, and W. Zabban, *Plating*, **39**, 267 (1952).
411. Setter, L. R., A. S. Goldin, and J. S. Nader, *Anal. Chem.*, **26**, 1304 (1954).
412. Setter, L. E., and A. S. Goldin, *J. Amer. Water Works Assoc.*, **48**, 1373 (1956).
413. Shain, I., "Stripping Analysis," in *Treatise on Analytical Chemistry*, Part I, Vol. 4, I. M. Kolthoff and P. J. Elving, Eds., Interscience, New York, 1959, p. 50.
414. Shain, I., and S. P. Perone, *Anal. Chem.*, **33**, 325 (1961).
415. Shapiro, J., *Science*, **133**, 2063 (1961).
416. Shapiro, J., *Science*, **133**, 3470 (1963).
417. Sherwood, R. M., and F. W. Chapman, *Anal. Chem.*, **27**, 88 (1955).
418. Shinn, M. B., *Ind. Eng. Chem.*, **13**, 33 (1941).
419. Sillen, L. G., *Advan. Chem. Ser.*, **67**, 45 (1967).
420. Simard, R. G., I. Hasegawa, W. Bandaruk, and C. E. Headington, *Anal. Chem.*, **23**, 1384 (1951).
421. Sirois, J. C., *Analyst*, **87**, 900 (1962).
422. Slavin, W., *Atomic Absorption Newsletter*, Perkin Elmer, Norwalk, Conn., **3**, 141 (1964).
423. Smith, R. S., and W. W. Walker, "Surveys of Liquid Wastes from Munitions Manufacturing," United States Public Health Service, Reprint 2508, Public Health Reports, 1965.
424. Smith, G. F., *Mixed Perchloric, Sulfuric and Phosphoric Acids and Their Applications in Analysis*, 2nd ed., G. Frederick Smith Chemical Company, Columbus, Ohio, 1942.
425. Smith, D., and J. Eichelberger, *J. Water Pollution Control Federation*, **37**, 77 (1965).
426. Soap and Detergent Association, Subcommittee on Biodegradation Test Methods, *J. Amer. Oil Chemists' Soc.*, **42**, 987 (1965).
427. Society for Analytical Chemistry, *Official, Standardized and Recommended Methods of Analysis*, S. C. Jolly, Ed., W. Heffer & Sons, Ltd., Cambridge, England, 1963.
428. Society of Public Analysts and Other Analytical Chemists, *Bibliography of Standard Tentative and Recommended or Recognized Methods of Analysis*, Society of Analytical Chemists Publication, London. 1951.
429. Sokolov, V. P., and K. A. Labashov, *Zavodsk. Lab.*, **28**, 285 (1962).
430. Soloway, S., and P. Rosen, *Anal. Chem.*, **29**, 1820 (1957).
431. Souliotis, A. G., R. P. Belkas, and A. P. Grimanis, *Analyst*, **92**, 300 (1967).
432. Spaulding, *J. Amer. Water Works Assoc.*, **34**, 277 (1942).
433. Stankovic, V., *Chem. Zvesti*, **16**, 683 (1962).
434. Stein, P. K., *U.S. Publ. Health Serv. Publ.*, **999-AP-15**, 65 (1964).
435. Stenger, V. A., and C. E. Van Hall, *Anal. Chem.*, **39**, 205 (1967).

436. Stephens, K.. *Limnol. Oceanog.*, **8**, 361 (1963).
437. Stewart, R. G., *Analyst*, **88**, 468 (1963).
438. Stojkovic, I., and H. Kukovec, *Textil*, **13**, 69 (1964).
439  Straub, C. P., *U.S. At. Energy Comm., Div. of Tech. Inform.*, **1966**, p. 88.
440. Streeter, H. W., and E. B. Phelps, *U.S. Public Health Bull.*, **146** (February 1925).
441. Strickland, J. D. H., and T. R. Parsons, *Bull. Fisheries Res. Board Can.*, **125**, 210 (1960).
442. Sullivan, J. P., "Determination of Dissolved Oxygen and Nitrogen in Sea Water by Gas Chromatography," Marine Science Department, Report No. 0-17-63, U.S. Naval Oceanographic Office, Washington, D.C., 1963.
443. Surak, J. G., M. F. Herman, and D. T. Haworgh, *Anal. Chem.*, **37**, 428 (1965).
444. Sweeney, W. A., and J. K. Foote, *J. Water Pollution Control Federation*, **36**, 14 (1964).
445. Swisher, R. D., *Soap Chem. Specialties*, **39**, 47, 57 (1963).
446. Swinnerton, J. W., V. J. Linnenbom, and C. H. Cheek, *Anal. Chem.*, **34**, 483 (1962).
447. Szekely, G., G. Raoz, and G. Traply, *Periodica Polytech.*, **10**, 231 (1966); *Chem. Abstr.*, **67**, 500 (1967).
448. Talvitie, N. A., and W. J. Garcia, *Anal. Chem.*, **37**, 851 (1965).
449. Taylor, A. E., and C. W. Miller, *J. Biol. Chem.*, **18**, 215 (1914).
450. Technical Association of the Pulp and Paper Industry, "Analysis of Industrial Process and Waste Waters," Tappi **12**, 32 (1955).
451. Technicon Instruments Corp., Bulletin, Technicon AutoAnalyzer, Channeey, New York.
452. Tenny, A. M., reprint from paper presented at Technicon Symposium on Automation in Analytical Chemistry, New York, October 19, 1966.
453. Thayer, L. C., and D. A. Robinson, *Proc. First Intern. Cong. Instr. Soc. Amer. 1st, Philadelphia*, **9**, 2 (1954).
454. Theriault, E. J., *U.S. Public Health Bull.*, **173** (July 1927).
455. Theriault, E. G., and P. D. McNamee, *Ind. Eng. Chem.*, **4**, 59 (1932).
456. Therous, F. R., E. F. Eldridge, and W. L. Mallmann, *Laboratory Manual for Chemical and Bacterial Analysis of Water and Sewage*, 3rd ed., McGraw-Hill, New York, 1943.
457. Thomas, H. A., Jr., *Sewage Works J.*, **9**, 425 (1937).
458. Thompson, M. E., and J. W. Ross, *Science*, **154**, 1643 (1966).
459. Tikhonov, M. K., and V. K. Zhavoronkina, *Soviet Oceanog.*, **3**, 22 (1960).
460. Todt, F., and R. Freier, Ger. Pat. No. 1-012-764 (1957).
461. Trubnik, E. H., and W. Rudolfs, *Proc. Ind. Waste Conf., 14th, Purdue Univ.*, **1948**, p. 266.
462. Trimmer, J. D., *Response of Physical Systems*, Wiley, New York, 1950.
463. Tschopp, E., and E. Tschopp, *Helv. Chim. Acta.*, **15**, 793 (1932).
464. Turekian, K. K., *U.S. At. Energy Comm.*, **2912-12**, 60 (1966).
465. Turski, Y. I., A. V. Masov, and L. E. Samolova, *Gaz. Prom.*, **4**, 20 (1959).
466. Tyler, C. P., and J. H. Karchmer, *Anal. Chem.*, **31**, 499 (1959).
467. Underwood, E. S., *Chem. Specialities Mfrs. Assoc. Proc.*, **1965**, 141.
468. Umbreit, W. W., R. H. Burris, and J. F Stanffer, *Manometric Techniques*, Burgess Publishing Co., Minneapolis, 1957.
469. Umland, F., and G. Z. Wunsch, *Anal. Chem.*, **213**, 186 (1965).

470. U.S. Department of the Interior, *Guidelines for Establishing Water Quality Standards for Interstate Waters—Under the Water Quality Act of 1965,* Public Law 89–234," Federal Water Pollution Control Administration, U.S. Department of the Interior (May 1966).

471. U.S. Public Health Service, National Water Quality Network, Annual Completion of Data, *Public Health Serv. Publ.,* 663 (1963).

472. U.S. Public Health Service and Ontario Department of Health, "A Study of Organic Contaminants in Boundary Waters Using Carbon Filter Techniques, Lake Huron—Lake Erie (1953–1955)," 1960.

473. United States Public Health Service, "Industrial Waste Guides: Supplement D," Ohio River Pollution Survey, *United States Public Health Service, Federal Security Agency,* 1942.

474. Van Hall, C. E., J. Safranko, and V. A. Stenger, *Anal. Chem.,* 35, 315 (1963).

475. Van Lopik, J. R., G. S. Rambie, and A. E. Pressman, *Proc. Symp. Water Quality Analysis, New York,* 1967, p. 7.

476. Van Slyke, D. D., and J. Folch, *J. Biol. Chem.,* 136, 509 (1940).

477. Van Slyke, D. D., *Anal. Chem.,* 26, 1706 (1954).

478. Velz, C. J., *Sewage Ind. Wastes,* 22, 666 (1950).

479. Voege, F. A., and H. L. Vanesche, *J. Amer. Water Works Assoc.,* 56, 1351 (1964).

480. Voth, J. L., *Anal. Chem.,* 35, 1957 (1963).

481. Walker, J. Q., *J. Gas Chromatograph,* 2, 46 (1964)

482. Ward, J. C., and R. W. Klippel, *Chem. Process. Mag.,* p. 69 (October 1965).

483. Warnick, S. L., and A. R. Gaufin, *J. Amer. Water Works Assoc.,* 57, 1023 (1965).

484. Water Pollution Control Federation, D.C., "Regulations of Sewer Use," Manual of Practice-3, 1955.

485. Weber, W. J., Jr., and R. H. Carlson, *J. Sanit. Eng. Div. Amer. Soc. Civil Eng.,* SA3, 140 (1965).

486. Weber, W. J., Jr., J. C. Morris, and W. Stumm, *Anal. Chem.,* 34, 1844 (1962).

487. Weber, W. J., Jr., and J. C. Morris, *J. Water Pollution Control Federation,* 36, 573 (1964).

488. Weber, W. J., Jr., and J. P. Gould, *Advan. in Chem. Ser.,* 60, 305 (1966).

489. Weiss, C. M., *Trans Amer. Fisheries Soc.,* 90, 143 (1961).

490. Weiss, C. M., and J. H. Gakstatter, *J. Water Pollution Control Federation,* 36, 240 (1964).

491. Weiss, C. M., J. D. Johnson, and B. Kwan, *J. Amer. Water Works Assoc.,* 55, 1367 (1963).

492. Weiker, W., *Anal. Chem.,* 185, 457 (1962).

493. Wells, W. N., P. W. Rohrbaugh, and G. A. Dotty, *Sewage Ind. Wastes,* 24, 212 (1952).

494. West, T. S., *Anal. Chim. Acta.,* 25, 405 (1961).

495. Westland, A. D., and R. R. Langford, *Anal. Chem.,* 33, 1306 (1961).

496. Wexler, A. S., *Anal. Chem.,* 35, 1936 (1963).

497. White, C. E., *Anal. Chem.,* 30, 729 (1958).

498. White, C. E., and J. Weissler, *Anal Chem.,* 36, 116R (1964).

499. Whitnack, G. C., *J. Electroanal. Chem.,* 2, 110 (1961).

500. Wickbold, R., *Z. Anal. Chem.,* 171, 81 (1959).

501. Wilcox, L. V., *J Amer. Water Works Assoc.,* 42, 775 (1950).

502. Wilks Scientific Corp. Publication, South Norwalk, Conn.
503. Williams, D. D., and R. R. Miller, *Anal. Chem.*, **34,** 657 (1962).
504. Wilson, R. F, *Limnol. Oceanog.*, **6,** 259 (1961).
505. Winefordner, J. D., and T. J. Vickers, *Anal. Chem.*, **36,** 161 (1964).
505a. Winkler, L. W., *Ber. Deut. Chem. Ges.*, **21,** 2843 (1888).
506. Winter, G. D., and A. Ferrari, "Automatic Wet Chemical Analysis as Applied to Pesticides Residues," in *Residue Rev. V*, W. Gunther, Ed., Springer-Verlag, Berlin, 1963.
507. Wise, W., B. F. Dodge, and H. Bliss, *Ind. Eng. Chem.*, **39,** 5, 632 (1947).
508. Witmer, A. W., and N. W. H. Addink, *Science Ind.*, **12,** 1 (1965).
509. Woodward, R. L., *Sewage Ind. Wastes*, **25,** 918 (1953).
510. World Health Organization, *World Health Organ. Tech. Rept. Ser.*, **173** (1959).
511. Yazhemskaya, V. Ya, *Gigiena i Sanit*, **29,** 58 (1964); *Chem. Abstr.*, **62,** 8305h (1965).
512. Young, J. C., and J. W. Clark, *J. Sanit. Eng. Div., Amer. Soc. Civil Engs.*, **SA1,** 43 (1965).
513. Yorko, N. A., and Z. A. Volkova, *Khim. Prom. Inform. Nauk-Tekh*, **2,** 77 (1964); *Chem. Abstr.*, **62,** 2611b (1965).
514. Yorko, N. A., and Z. A. Volkova, *Khim. Prom. Inform. Nauk-Tekh*, **2,** 77 (1964).
515. Zhavoronkina, T. K., *Soviet Oceanog.*, **3,** 27 (1960).